Degradation Processes in Nanostructured Materials

MATERIALS RESEARCH SOCIETY
SYMPOSIUM PROCEEDINGS VOLUME 887

Degradation Processes in Nanostructured Materials

Symposium held November 28–December 1, 2005, Boston, Massachusetts, U.S.A.

EDITORS:

Mircea Chipara
Indiana University
Bloomington, Indiana, U.S.A.

Orazio Puglisi
University of Catania
Catania, Italy

Ralph Skomski
University of Nebraska-Lincoln
Lincoln, Nebraska, U.S.A.

Frank R. Jones
University of Sheffield
Sheffield, United Kingdom

Benjamin S. Hsiao
State University of New York-Stony Brook
Stony Brook, New York, U.S.A.

Materials Research Society
Warrendale, Pennsylvania

CAMBRIDGE
UNIVERSITY PRESS

University Printing House, Cambridge CB2 8BS, United Kingdom

One Liberty Plaza, 20th Floor, New York, NY 10006, USA

477 Williamstown Road, Port Melbourne, VIC 3207, Australia

314-321, 3rd Floor, Plot 3, Splendor Forum, Jasola District Centre, New Delhi - 110025, India

79 Anson Road, #06-04/06, Singapore 079906

Cambridge University Press is part of the University of Cambridge.

It furthers the University's mission by disseminating knowledge in the pursuit of education, learning and research at the highest international levels of excellence.

www.cambridge.org
Information on this title: www.cambridge.org/9781558998414

Materials Research Society
506 Keystone Drive, Warrendale, PA 15086
http://www.mrs.org

First published 2006
First paperback edition 2012

Single article reprints from this publication are available through University Microfilms Inc., 300 North Zeeb Road, Ann Arbor, MI 48106

CODEN: MRSPDH

A catalogue record for this publication is available from the British Library

ISBN 978-1-558-99841-4 Hardback
ISBN 978-1-107-40891-3 Paperback

CONTENTS

*Invited Paper

*Invited Paper

DEGRADATION PROCESSES AT
NANOMETER SCALE

*Invited Paper

PREFACE

The unifying theme of Symposium Q, "Degradation Processes in Nanostructured Materials," held November 28–December 1 at the 2005 MRS Fall Meeting in Boston, Massachusetts, was natural aging and artificial degradation in a broad variety of nanostructured materials. The considered structures—including nanotubes, nanofibers, thin films, nanoparticles and composites—were made from various types of substances, such as elements (C, Si, Sn), oxides, metals, and polymers. In addition, some papers dealt with specific structural features, such as surfaces, interfaces, and grain boundaries. The nanoscale character of the materials, as compared to the micron-size structuring of many traditional materials, leads to quantitative changes and to new physical phenomena. The reason is that the involved structural length scales are comparable to physical length scales such as gyration radii in polymers and domain-wall widths in magnetic materials.

The interdisciplinary character of the symposium was manifested by the breadth of the phenomena and applications which were considered; from electronic, optical, structural, and magnetic nanomaterials to space, chemical, biological and medical applications. A key theme in many invited and contributed talks, was the interplay between structural degradation and the time dependence of material properties and merit indices. Aside from natural aging, there was emphasis on the time dependence associated with artificial treatments involving electrical currents, annealing, and irradiation by various sources, such as gamma radiation and ions.

Some papers focused on nanoscale thermal degradation, corrosion, flammability, embrittling, and fatigue. Among the irradiation effects, which are important not only in space applications, were changes in mechanical and functional properties arising from gamma and ion irradiation.

Finally, we are pleased to share the success of this symposium with the authors and participants and to acknowledge the continuous help, support, and understanding of the Materials Research Society staff.

Mircea Chipara
Orazio Puglisi
Ralph Skomski
Frank R. Jones
Benjamin S. Hsiao

February 2006

MATERIALS RESEARCH SOCIETY SYMPOSIUM PROCEEDINGS

MATERIALS RESEARCH SOCIETY SYMPOSIUM PROCEEDINGS

Prior Materials Research Society Symposium Proceedings available by contacting Materials Research Society

Degradation Processes in
Carbon Nanotubes

Mater. Res. Soc. Symp. Proc. Vol. 887 © 2006 Materials Research Society   0887-Q01-07

Photo-oxidation of Single-walled Carbon Nanotubes.

B. Parekh, T. Debies[1], C. M. Evans[2], B. J. Landi[2], R. P. Raffaelle[2] and G. A. Takacs[*]

Department of Chemistry, Center for Materials Science and Engineering,
Rochester Institute of Technology, Rochester, NY, 14623, U.S.A.
[1]Xerox Corporation, Webster, NY 14580, U.S.A.
[2]NanoPower Research Laboratories, RIT, Rochester, NY, 14623, U.S.A.

ABSTRACT

UV- and vacuum UV-assisted photo-oxidation of single-walled carbon nanotube (SWNT) paper occurred with: (1) atmospheric oxygen pressure using low-pressure Hg lamps (λ = 253.7 and 184.9 nm), (2) low oxygen pressure employing emission downstream from an Ar microwave plasma (λ = 106.7 and 104.8 nm), and (3) high pressures of He in a rotating d.c. arc that was designed to produce a spectral continuum from He excimers (λ = 58 - 110 nm). The photo-oxidized materials were characterized by x-ray photoelectron spectroscopy (XPS), scanning electron microscopy (SEM) and Raman spectroscopy. UV photo-oxidation was demonstrated to be a controlled dry procedure for introducing oxygenated functional groups (C-O-C, C=O, O-C=O and O=C-O-C=O) on SWNTs.

INTRODUCTION

Previously, a number of methods have been used to oxidize carbon nanotubes (CNTs). Gas-phase oxidation in air at around 750 °C leads to over oxidation often removing or severely damaging the CNTs in addition to reacting with amorphous carbon [1]. Liquid-phase oxidation, involving nitric and/or sulfuric acids, is mild and slow and produces high yields of oxidized CNTs [2] having a mixture of –C-O-, -C=O and -COO- functional groups as observed by x-ray photoelectron spectroscopy (XPS) [3]. Ozonolysis, both in the liquid [4] and gas phase [5 - 7], introduces oxygenated functionalities directly onto the sidewalls and not simply at the end caps or isolated defects [8]. Use of UV/O$_3$ in air with single-walled carbon nanotubes (SWNTs), shows by TGA and Raman analysis that ca. 5% of the carbons were functionalized and rapid initial oxidation occurs within 1 h of treatment which stops after 3 h probably due to exhaustion of active surface sites [9]. The UV/O$_3$ method is a dry technique that produces ozone *in situ* and, therefore, does not involve transport of reactive ozone to the reaction chamber from an electric ozonizer or liquid waste from solvent or a hydrolysis step. XPS analysis of the top 2 – 5 nm of the surface of multi-walled carbon nanotubes (MWNTs) shows that gas-phase UV and vacuum UV (VUV) photo-oxidation achieves oxygen concentrations up to 7.5 and 9.5 at% after treatment times of 4 and 2 h, respectively [10]. Curve fitting of the XPS C1s spectra reveals mainly the C-O-C functional group with the presence of C=O, O-C=O and O=C-O-C=O moieties [10].

In the present work, purified laser generated SWNT papers are photo-oxidized with UV and VUV radiation and studied with XPS, Raman and SEM microscopy.

EXPERIMENTAL DETAILS

Synthesis of SWNTs was performed using an Alexandrite laser vaporization process, previously described in detail [11]. In summary, a graphite (1-2 μm) target was pressed at 20,000 psi and contained 3% w/w Ni (sub-micron) and 3% w/w Co (1-2 μm). The reaction furnace temperature was maintained at 1150 °C, with a chamber pressure of 700 torr under 100 sccm flowing Ar. The raw SWNT soot was purified by conventional acid reflux. The ratio of materials for refluxing was: 75 mg raw soot added to the acid solution (50 mL H_2O, 12 mL concentrated HNO_3 (69-70 %), and 5 mL concentrated HCl (36.5-38.0 %)). The solution was brought to reflux at 120 °C for 14 h. The reflux solution was filtered over a 1 μm PTFE membrane filter with copious distilled H_2O to form the SWNT paper. The acid filtrate was discarded and subsequent washes (3x) with 50 mL acetone and 10 mL distilled H_2O removed functionalized carbon impurities until the filtrate was clear. To complete the purification, the resulting paper from acid-reflux was thermally oxidized at 550 °C in the muffle furnace under stagnant air for 1 h.

Photo-oxidation of SWNT paper was studied using: (1) atmospheric oxygen pressure with low-pressure Hg lamps (λ = 253.7 and 184.9 nm) [12], (2) low oxygen pressure with emission downstream from an Ar microwave (MW) plasma (λ = 106.7 and 104.8 nm) [13], and (3) high pressures of He in a rotating d.c. arc that was designed to produce a spectral continuum from He excimers, He_2^*, (λ = 58-110 nm) [14 – 16].

A Physical Electronics Model 5800 XPS system provided elemental, chemical state and quantitative analyses of the top 2 – 5 nm of a sample's surface at an angle of 45° between the sample and analyzer.

A Hitachi S-900 field emission SEM was used to analyze the surface morphology of the photo-oxidized SWNTs. The samples were applied directly to the brass stub using silver paint. The instrument operated at an accelerating voltage of 2kV and magnifications ranging from 5 – 70 kX.

Raman spectroscopy was performed at room temperature using a JY-Horiba Labram spectrophotometer with excitation energy of 1.96 eV. Sample spectra were obtained from 50 to 2800 cm^{-1} using an incident beam attenuation filter to eliminate localized heating and subsequent sample decomposition.

DISCUSSION

The oxygen concentration of UV photo-oxidized SWNTs increased rapidly initially with exposure time, but after 45 min, it slowed eventually reaching a value of 25.5 at% after 2 h of treatment time (Fig. 1). These nanoscale XPS measurements of the level of oxidation are larger than that reported for the % carbon functionalized as detected by microscale TGA data for UV/O_3 treatment in air of HiPco SWNT bucky paper (ca. 5%) [9] and infrared measurements of gaseous CO and CO_2 released following ozone treatment of closed end SWNTs prepared by laser ablation of graphite targets (5.5 ±2.5 %) [7]. XPS detection of the oxidation of MWNTs have been reported to have values up to 5 – 7.5 at% for UV photo-oxidation [10], KMnO$_4$/H$_2$SO$_4$ oxidation [17] and oxidation with a nitric/sulfuric acid mixture [3].

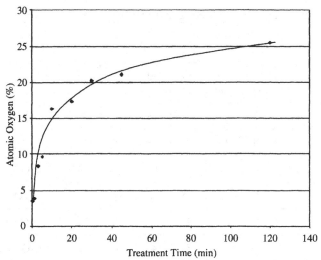

Figure 1. % O as a function of exposure time for UV photo-oxidized SWNTs

Figure 2 illustrates the overlapped C 1s XPS spectra obtained for the SWNTs exposed to different UV photo-oxidation treatment times. The spectra show a significant increase in the intensity of high binding energy peaks (~290-286 eV) associated with carbon-oxygen bonding. The peaks are assigned to species reported in the literature [18] and the areas are tabulated in Table I. Longer exposure time results in higher percentages of oxygen-containing functionalities as well as a reduction in C-C bonding.

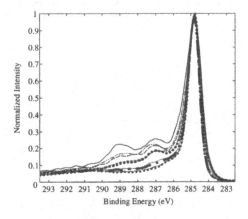

Figure 2. C 1s XPS Spectra for UV photo-oxidized SWNTs as a function of treatment time from 0 (....) to 120 min. (___).

Table I. Results of the Curve Fittings for C 1s Peaks of the UV Photo-oxidized SWNTs

B. E (eV)	Peak Area %									Assignment
	0 min	1 min	3 min	5 min	10 min	20 min	30 min	45 min	2 h	
284.8	51	51	45	43	52	51	46	45	39	C-C sp^2
285.2	28	24	25	28	15	17	14	15	16	C-C sp^3
286.0	8	11	12	14	9	9	12	11	14	C-O-C
287.3	4	4	6	7	11	10	12	12	11	C=O
288.6	3	3	4	3	7	7	9	10	11	-O-C=O
289.8	3	3	4	2	3	3	4	4	6	O=C-O-C=O
291.0	3	4	4	3	3	3	3	3	3	Energy Loss

Changes in the surface morphology were monitored using SEM and are shown in Fig. 3. For treatment times up to 45 min, there was little variation from the untreated SWNTs. However, beginning at ca. 45 min, changes became apparent in the bundle size of the SWNTs and the concentration of amorphous carbon. Additionally, changes in the Raman radial breathing mode (RBM) after 45 min suggest that the diameter distribution was altered, which is consistent with the presence of amorphous carbon shown in the SEM images. These results suggest that at longer exposure times the SWNTs may actually be consumed while at shorter times (< 45 min) there is controlled functionalization.

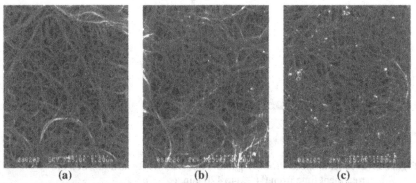

(a)   (b)   (c)

Figure 3. SEMs for SWNTs: treated for (a) 20 min, (b) 45 min and (c) 120 min of UV photo-oxidation. All SEMs are for a magnification =25 kX, sizing =1.20 μm.

Other photo-oxidation techniques, having different photon energies, also are being explored. Figure 4 shows the Raman spectra for a control sample and samples treated for 3 h with UV photo-oxidation (4.9 and 6.7 eV), 2.5 h of VUV photo-oxidation downstream from the MW Ar plasma (11.6 and 11.8 eV) and 5 min with the rotating He arc (11.3 – 21.4 eV) which resulted in O at% values of 4.7, 29.2, 15.8 and 14.0, respectively. The Raman spectra consist of a distinctive pair of broadband peaks at ~1580 (G-band) and ~1340 cm^{-1} (D-band) corresponding to sp^2 and defect-induced carbon stretching modes, respectively [19]. The D-band is caused by symmetry-breaking defects such as sp^3-hybidized carbon atoms, as well as, hetero-atoms, vacancies, heptagon-pentagon pairs, kinks or the presence of impurities [19]. The data shows that all three techniques of photo-oxidation increased the ratios of the intensities of the D:G bands as compared to the untreated sample with the largest increase for the sample treated using the UV low pressure Hg lamps.

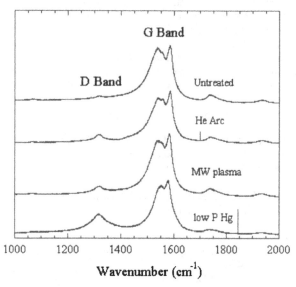

Figure 4. Raman spectra taken at 1.96 eV for untreated SWNTs and SWNTs treated with He arc discharge, microwave Ar plasma and UV low pressure Hg lamps.

CONCLUSIONS

Dry techniques, involving UV- and VUV- assisted photo-oxidation, successfully introduced oxygenated functional groups (C-O-C, C=O, O-C=O and O=C-O-C=O) on SWNT papers as observed by XPS spectroscopy of the top 2 -5 nm of the sample's surface. SEM and Raman spectroscopy provided complementary results which helped show that controlled functionalization occurred at short treatment times with UV photo-oxidation, while at longer treatment times, the bundles were eroded and the diameter distribution was altered. Higher levels of oxidation were achieved with SWNTs than MWNTs [10].

*To whom correspondence should be addressed. Phone 585-475-2047. Fax: 585-475-7800. E-mail address: gatsch@rit.edu.

REFERENCES

1. T. Ebbesen, H. Hiura, H. Fujita and K. Tanigaki, *Nature* **367**, 519 (1994).
2. H. Hiura, T. W. Ebbesen and K. Tanigaki, *Adv. Mater.* **7**, 275 (1995).
3. Y. Xing, L. Liang, C. C. Chusuei and R. V. Hull, *Langmuir* **21**, 4185 (2005).
4. S. Banerjee and S. S. Wong, *J. Phys. Chem. B* **106**, 12144 (2002).
5. X. Lu, L. Zhang, X. Xu, N. Wang and Q. Zhang, *J. Phys. Chem.* **B106**, 2136 (2002).
6. O. Byl, J. Liu and J. T. Yates, Jr., *Langmuir* **21**, 4200 (2005).
7. D. B. Mawhinney, V. Naumenko, A. Kuznetsova, J. T. Yates, Jr., J. Liu and R. E. Smalley, *Chem. Phys. Lett.* **324**, 213 (2000).
8. D. B. Mawhinney, V. Naumenko, A. Kuznetsova, J. T. Yates, Jr., J. Liu and R. E. Smalley, *J. Am. Chem, Soc.* **122**, 2383 (2000).
9. L. Cai, J. L. Bahr, Y. Yao and J. M. Tour, *Chem. Mater.* **14**, 4235 (2002).
10. B. Parekh, T. Debies, P. Knight, K. S. V. Santhanam and G. A. Takacs, *Mat. Res. Soc. Sym. Proc.* (Degradation Processes in Nanostructured Materials) 2005 MRS Fall Meeting, this issue (2006).
11. B. J. Landi, R. R. Raffaelle, S. L. Castro and S. G. Bailey, *Prog. Photovolt: Res. Appl.* **13**, 165 (2005).
12. U. Sener, A. Entenberg, B. Kahn, F. D. Egitto, L. J. Matienzo, T. Debies and G. A. Takacs, in: *Polyimides and Other High Temperature Polymers: Synthesis, Characterization and Applications*, K. L. Mittal (Ed.), Vol. **3**, 535 – 552, VSP, Utrecht (2005).
13. W. Dasilva, A. Entenberg, B. Kahn, T. Debies and G. A. Takacs, *Mater. Res . Soc. Symp. Proc.* **851** (Materials for Space Applications), 407 – 412 (2005); *J. Adhesion Sci. Technol.* **18**, 1465 – 1481 (2004).
14. G. A. Takacs, V. Vukanovic, F. Egitto, L. J. Matienzo, F. Emmi, D. Tracy and J. X. Chen, *10th IUPAC International Symp. on Plasma Chemistry (ISPC-10)*, Bochum, Germany, **1.4-29**, p. 1-6 (1991); G. A. Takacs, V. Vukanovic, D. Tracy, J. X. Chen, F. D. Egitto, L. J. Matienzo and F. Emmi, *Poly. Degrad. Stabil.* **40**, 73 (1993).
15. J. X. Chen, D. Tracy, S. Zheng, L. Xiaolu, S. Brown, W. VanDerveer, A. Entenberg, V. Vukanovic, G. A. Takacs, F. D. Egitto, L. J. Matienzo and F. Emmi, *Poly. Degrad. Stabil.* **79**, 399 (2003).
16. S. Zheng, A. Entenberg, G. A. Takacs, F. D. Egitto and L. J. Matienzo, *J. Adhesion Sci. Technol.* **17**, 1741 (2003).
17. S. Banerjee and S. S. Wong, *J. Am. Chem. Soc.*, **125**, 10342 (2003).
18. G. Beamson and D. Briggs, *High Resolution XPS of Organic Polymers*, John Wiley & Sons Ltd., Chichester, West Sussex, England, p.158 (1992).
19. A. Jorio, M. A. Pimenta, A. G. Souza Filho, R. Saito, G. Dresselhaus and M. S. Dresselhaus, *New Journal of Physics* **5**, 139.1 (2003).

Mater. Res. Soc. Symp. Proc. Vol. 887 © 2006 Materials Research Society   0887-Q06-01

Photo-oxidation of Multiwalled Carbon Nanotubes

B. Parekh, T. Debies[1], P. Knight, K. S. V. Santhanam and G. A. Takacs
Department of Chemistry, Center for Materials Science and Engineering,
Rochester Institute of Technology, Rochester, NY 14623, U.S.A.
[1]Xerox Corporation, Webster, NY 14580, U.S.A.

ABSTRACT

MWNTs were photo-oxidized in quartz boats using low-pressure Hg lamps and emission downstream from Ar microwave plasmas which are primarily atomic line sources of 253.7 and 184.9 nm UV, and 106.7 and 104.8 nm vacuum UV (VUV) radiation, respectively. X-ray photoelectron spectroscopy (XPS) showed rapid oxidation during the first hour of UV treatment and then an increase that was directly proportional to the time of treatment up to 4 h where the oxygen concentration was 7.5 at%. VUV photo-oxidation resulted in an oxygen concentration up to 9.5 at% with exposure time for the initial 2 h of treatment. Beyond 2 h, the oxygen concentration decreased with VUV exposure due to a larger rate of de-oxidation than oxidation at the surface. Curve fitting of the XPS C1s spectra revealed mainly the C-O-C functional group with the presence of C=O, O-C=O and O=C-O-C=O moieties. SEM micrographs showed no apparent effect on the structure or appearance of the MWNTs as expected for surface modification. Gas phase photo-oxidation effectively functionalizes MWNTs without resulting in liquid waste from the traditional method of acidic oxidation.

INTRODUCTION

Gas-phase photo-oxidation of MWNTs were studied at 298 K with wavelengths from low pressure Hg lamps ($\lambda = 184.9$ nm) and excited Ar atoms ($\lambda = 106.7$ and 104.8 nm) downstream from a MW plasma which have sufficient energy to photo-dissociate gaseous oxygen and result in chemical modification of the surface. Earlier work reported in the literature focused on UV wavelengths (240 and 253.7 nm) photo-absorbed by the carbon nanotubes (CNTs) that did not change the surface composition but caused photo-desorption of O_2 [1] and changed the thermoelectric power [2, 3] while a UV/O_3 generator in air formed oxygenated functional groups on single-walled carbon nanotubes (SWNTs) [4].

EXPERIMENTAL DETAILS

The MWNTs, which were obtained in a powder form from Helix Material Solution, were produced by an electric arc method and characterized as > 95% pure and having a range of diameters and lengths from 60 - 100 nm and 0.5 - 40 μm, respectively. The MWNTs were studied by placing the powder in a well of a quartz block. The UV and VUV experimental conditions and equipment are as reported in refs. [5] and [6].

The samples were analyzed with a Physical Electronics Model 5800 XPS that analyses the top 2 – 5 nanometers of a sample's surface with a take-off angle of 45° between the sample and analyzer. The quartz boat was positioned in a stainless steel sample holder so that the nanotubes were flush with the surface of the sample holder.

The quantitative analyses are precise to within 5% relative for major constituents and 10% relative for minor constituents. The samples were irradiated with monochromatic Al Kα radiation (1486 eV) and charge neutralized with a flood of low energy electrons from a BaO field emission charge neutralizer which minimized radiation damage.

An Amray 3300 field emission SEM was used to analyze the surface morphology of the photo-oxidized MWNT powder in the quartz boats that were placed onto carbon adhesive tabs on Al stubs and coated with a thin conductive Au film of ca. 20 nm.

DISCUSSION

The MWNTs contained only carbon and a trace amount of oxygen (ca. 2 at%). Photo-oxidation of MWNTs for 3.5 h with 253.7 nm radiation resulted in little effect on the chemical bonding as observed by the XPS C 1s and O 1s spectra consistent with the earlier work using the UV wavelengths of 240 and 253.7 nm [1-3].

When 253.7 and 184.9 nm radiation lamps were used, increasing oxidation of the MWNTs occurred as a function of exposure time up to 4 h where the O concentration was 7.5 at%. The data, which is plotted in Fig. 1, shows a linear relationship from 1 to 4 h of treatment with a correlation coefficient of r-squared = 0.96.

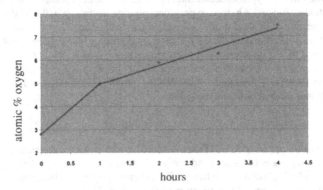

Figure 1. % O for MWNT samples as a function of exposure time to UV photo-oxidation.

The nanotubes exposed to VUV photo-oxidation contained an oxygen concentration up to 9.5 at% at 2 h of exposure time. As with the UV photo-oxidation results, most of the VUV photo-oxidation occurred within 1 h of treatment time. Beyond 2 h, the concentration of oxygen on the VUV photo-oxidized surface decreased and was equal to 4.2 at % at 4 h. SEM micrographs revealed no apparent effect on the structure or appearance of the UV (25kX, 1μm sizing bar) and VUV photo-oxidized MWNTs (100 kX, 300 nm sizing bar).

The C 1s spectra obtained from untreated samples were in good agreement with results reported in the literature [7, 8] and differed only slightly from the spectra obtained for the UV and VUV photo-oxidized nanotubes due to the low concentration of oxygen introduced by exposure. Figure 2 shows the C 1s spectrum for MWNTs treated for 3 h of

UV photo-oxidation. The curve fitting, which was employed to interpret the C 1s spectra, was similar to that used by ref. [7]. The peaks at 284.8 and 285.2 eV are due to sp^2- and sp^3-hybidized carbon, respectively. To obtain consistency with the oxygen concentration measured post treatment, the peaks at 286.3, 287.5 and 288.9 eV were assigned to the C-O-C, C=O, O-C=O groups, respectively. The peak at 290.0 eV is difficult to assign to the nanotubes and is attributed to the O=C-O-C=O moiety. The peak assignments and areas are shown in Table I for the UV and VUV photo-oxidized samples. The differences in binding energies (B.E.) are due to errors associated with the curve fittings.

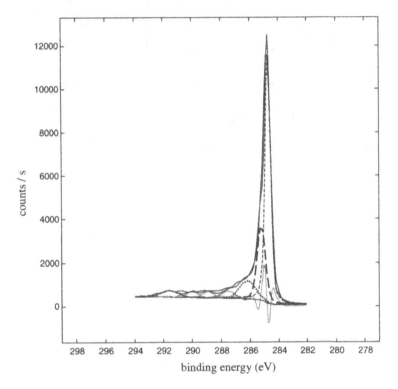

Figure 2. Curve fitting of C 1s spectrum for MWNTs treated for 3 h of UV photo-oxidation

The untreated sample yielded a weak and broad O 1s spectrum. The treated nanotubes gave similar O 1s spectra. Because the O 1s spectrum has no energy resolved peak structure, it is difficult to know how many peaks are present and their exact energy location. Analogous to the curve fitting of the O 1s peaks for the study of oxidation of MWNTs with HNO_3 and H_2SO_4 [8], Fig. 2 shows the O 1s spectrum for MWNTs treated for 4 h with UV photo-oxidation and curve fit into a mixture of -C=O,-C-O, -COO- and energy loss entities with increasing binding energies, respectively.

Table I. XPS Results of the C 1s Curve Fittings for MWNTs Treated for 3 h with UV or 2 h VUV Photo-oxidation

B.E. (eV) UV	Peak Area % UV	B.E. (eV) VUV	Peak Area % VUV	Assignment
284.8	55.7	284.8	61.6	C-C sp^2
285.2	22.8	285.1	14.2	C-C sp^3
286.3	9.9	286.2	11.2	C-O-C
287.5	3.4	287.2	4.4	C=O
288.9	2.4	288.6	4.0	O-C=O
290.0	2.1	289.8	2.1	O=C-O-C=O
291.2	3.7	291.6	2.4	Energy Loss

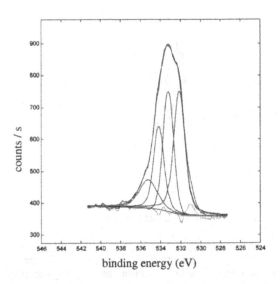

Figure 3. Curve fitting for O 1s XPS spectrum of MWNTs treated with UV photo-oxidation for 4 h.

Both oxygen atoms [9] and ozone [10], formed following photo-dissociation of oxygen molecules, oxidize by adding across unsaturated bonds, like aromatic groups. Chemisorbed oxygen atoms weaken the zigzag C-C bonds of CNTs to replace them with C-O-C bridge bonds [11]. Initially, the addition of ozone produces a primary ozonide that undergoes bond cleavage to form carbonyl compounds and a Criegee intermediate that may decompose into an ester group (O-C=O) and the potential release of CO_2 [10, 12].

Gas phase ozonolysis of SWNTs at 298 K has previously indicated the production of ester groups, carbonyl (C=O), and C=C double bonds by FT-IR analysis [13]. In addition, the formation of CO_2 (g) and CO (g) suggested the existence of a carbon removal process [13, 14]. Sonication in HNO_3 and H_2SO_4 effectively functionalizes MWNTs to form a mixture of –C-O-, -C=O and –COO- groups as observed by XPS with the bulk of the oxidation occurring between 1 and 2 h and the –C-O- group taking up the greater of the population [8]. Similar results were obtained in this gas phase UV photo-oxidation study, with the majority of the oxidation occurring within the first hour and primarily forming the C-O-C functional group. The use of an UV/O_3 generator in air showed by TGA and Raman analysis ca. 5% of the O_3-SWNT carbons were functionalized and that rapid initial oxidation occurred within 1 h of treatment which stopped after 3 h due to exhaustion of active surface sites [4]. The MWNTs in this UV study showed up to 7.5 at% oxygen with available surface sites from 1 to 4 h (Fig. 1). XPS analysis of MWNTs, that were oxidized with acids, have yielded similar levels of oxidation (5-6 at% [12(a)], ca. 6 at% [8]).

Beyond 2 h of treatment time with VUV photo-oxidation, the O at% on the surface of the MWNTs begins to decrease. Lim et al [15] has reported that when MWNTs at 700 °C are exposed to oxygen atoms oxygen is desorbed from the surface through the formation of CO and CO_2. The reaction of atomic oxygen with the VUV photo-exposed oxygenated surface may also result in such de-oxidation processes.

An additional mechanism to be considered is photo-excitation of the MWNTs. The energies associated with the UV (6.7 eV) and VUV (11.6 and 11.8 eV) photons are greater than the bond energies for C-C (~3 eV) and weaker unsaturated aromatic bonds [16]. Therefore, UV and VUV photo-absorption of the MWNTs may initiate photo-dissociation of bonds or cause photo-ionization. C_{60} may be UV-sensitive in the absence of oxygen [17] and has been reported to undergo photoelectron detachment using ArF (6.4 eV) excimer radiation [18]. The capture of low energy electrons by O_2 to form the reactive O_2^- species has been proposed to be involved in the photo-induced oxidation of multilayer films of fullerene C_{60} at 20 K [19].

CONCLUSIONS

Treatment of MWNTs with gas phase UV and VUV photo-oxidation resulted in surface oxidation at levels similar to previous work involving acidic oxidation. Curve fitting of the XPS C1s spectra showed the largest contribution due to the C-O-C moiety and the presence of C=O, O-C=O and O=C-O-C=O groups. SEM micrographs indicated no apparent effect on the appearance of the surface modified MWNTs.

ACKNOWLEDGEMENTS

The authors are thankful to G. Thompson from Xerox Corp. for performing the SEM studies.

REFERENCES

1. R. J. Chen, N. R. Franklin, J. Kong, J. Cao, T. W. Tombler, Y. Zhang and H. Dai, *Appl. Phys. Lett.* **79**, 2258 (2001).
2. T. Savage, S. Bhattacharya, B. Sadanadan, J. Gaillard, ''. M. Tritt, Y-P. Sun, Y. Wu, S. Nayak, R. Car, N. Marzari, P. M. Ajayan and A. M. Rao, *J. Phys.: Condens. Matter* **15**, 5915 (2003).
3. M. Grujicic, G. Cao, A. M. Rao, T. M. Tritt and S. Nayak, *Appl. Surf. Sci.* **214**, 289 (2003).
4. L. Cai, J. L. Bahr, Y. Yao and J. M. Tour, *Chem. Mater* **14**, 4235 (2002).
5. U. Sener, A. Entenberg, B. Kahn, F. D. Egitto, L. J. Matienzo, T. Debies and G. A. Takacs, in: *Polyimides and Other High Temperature Polymers: Synthesis, Characterization and Applications*, K. L. Mittal (Ed.), Vol. 3, 535 – 552, VSP, Utrecht (2005).
6. W. Dasilva, A. Entenberg, B. Kahn, T. Debies and G. A. Takacs, *J. Adhesion Sci. Technol.* **18**, 1465 – 1481 (2004).
7. Z. Konya, I. Vesselenyi, K. Niesz, A. Kukovecz, A. Demortier, A. Fonseca, J. Delhalle, Z. Mekhalif, J. B. Nagy, A. A. Koos, Z. Osvath, A. Kocsonya, L. P. Biro and I. Kiricsi, *Chem. Phys. Lett.* **360**, 429 (2002).
8. Y. Xing, L. Liang, C. C. Chusuei and R. V. Hull, *Langmuir* **21**, 4185 (2005).
9. B. J. Finlayson-Pitts and J. N. Pitts, Atmospheric Chemistry, p. 459, Wiley & Sons, New York (1986).
10. B. J. Finlayson-Pitts and J. N. Pitts, Chemistry of the Upper and Lower Atmosphere, Academic Press, New York (2000).
11. S. Dag, O. Gulseren, T. Yildirim and S. Ciraci, *Phys. Rev.* **B67**, 165424 (2003).
12. (a) S. Banerjee and S. S. Wong, *J. Phys. Chem.* **B106**, 12144 (2002).
    (b) X. Lu, L. Zhang, X. Xu, N. Wang and Q. Zhang, *J. Phys. Chem.* **B106**, 2136 (2002).
13. D. B. Mawhinney, V. Naumenko, A. Kuznetsova, J. T. Yates, Jr., J. Liu and R. E. Smalley, *J. Am. Chem. Soc.* **122**, 2383 (2000).
14. D. B. Mawhinney, V. Naumenko, A. Kuznetsova, J. T. Yates, Jr., J. Liu and R. E. Smalley, *Chem. Phys. Lett.* **324**, 213 (2000).
15. S. C. Lim, H. J. Jung, D. J. Bae, E. K. Suh, Y. M. Shin, K. H. An, Y. H. Lee and J. Choi, *Proc. Sixth Appl. Diamomd Conf. / Second Frontier Carbon Technol. Joint Conf. (ADC/FCT 2001)*, Y. Tzeng, K. Miyoshi, M. Yoshikawa, M, Murakawa, Y. Koga, K. Kobashi and G. A. J. Amaratunga (Eds.), 759 (2001).
16. T. L. Cottrell, *The Strengths of Chemical Bonds*, 2nd ed., Butterworths, Washington, DC (1958).
17. R. Taylor, J. P. Parsons, A. G. Avent, S. P. Rannard, T J. Dennis, J. P. Hare, H. W. Kroto and D. R. M. Walton, *Nature* **351**, 277 (1991).
18. R. E. Haufler, L-S. Wang, L. P. F. Chibante, C. Jin, J. Conceicao, Y. Chai and R. E. Smalley, *Chem. Phys. Lett.* **179**, 449 (1991).
19. G. H. Kroll, P. J. Benning, Y. Chen, T. R. Ohno, J. H. Weaver, L. P. F. Chibante and R. E. Smalley, *Chem. Phys. Lett.* **181**, 112 (1991).

Mater. Res. Soc. Symp. Proc. Vol. 887 © 2006 Materials Research Society   0887-Q01-03

Raman and Electron Spin Resonance in as Prepared and Thermal Annealed Cluster Beam Deposited Carbon Thin Films

Giuseppe Compagnini, Orazio Puglisi, Mircea Chipara[1]

Dipartimento di Scienze Chimiche, Universita di Catania
[1]Indiana University Cyclotron Facility 2401 Milo B Sampson Lane, Bloomington, Indiana, IN 47408.

ABSTRACT

Nanostructured amorphous carbon thin films deposited by a Low Energy Carbon Cluster Beam Deposition technique have been characterized by Raman spectroscopy and Electron Spin Resonance (ESR). This study focuses on the correlation between carbon cluster structure and sp^2 related defects. The ESR spectrum of as prepared and thermally annealed carbon clusters is a single line located near the free electron g-factor. Resonance lines were accurately fitted by a single line symmetric and narrow Lorentzian line. The resonance line parameters (line position, line amplitude, and line width) were found to be extremely sensitive to the thermal treatment of the as deposited samples. The paramagnetic centres are randomly distributed and correlated with the nanosized nature of the investigated system. The absence of the resonance line asymmetry, typical for the ESR spectra of conducting carbaceneous materials, confirms the small size of carbon nanoclusters and indicates poor electrical contacts between them. The resonance line parameters (position, intensity, and width) are not affected by the orientation of the carbon cluster film relative to the direction of the external magnetic field. The evolution of the ESR signal upon thermal treatments is sensitive to the increase of the sp^2 average domain size as revealed by Raman spectroscopy. The Raman spectrum of as obtained and thermally annealed carbon clusters, in the domain 1000 cm^{-1} to 2000 cm^{-1} shows two Lorentzian lines located at 1300 cm^{-1} and 1550 cm^{-1} representing the D and G bands. The position and amplitude of these lines were found to be affected by thermal annealing.

INTRODUCTION

Thin carbon films exhibit a wide range of structures such as 3-dimensional graphitic domains, quasi-two dimensional graphitic like structures (highly oriented pyrolytic graphite), amorphous carbon structures, and eventually fullerenes and extremely short nanotubes. Raman spectroscopy is an excellent tool in the investigation of thin carbon films. Raman spectroscopy provides detailed information regarding the vibration of carbaceneous materials. Raman signals [2] located between

1000 cm^{-1} and 2000 cm^{-1} are mainly due to the presence of sp^2 components in the film. This is the most relevant part of the spectrum and a huge amount of papers have been written [1-3] to account for the correlation between these features and the atomic and electronic structure of the investigated sample. The region is generally considered as composed by two features located around 1330 cm^{-1} to 1380 cm^{-1} (D line) and 1540 cm^{-1} to 1600 cm^{-1} (G line). The position, width and intensity of these two lines are strongly dependent on the size of the obtained carbon particles and on their electronic structure.

Electron spin resonance was frequently used to investigate carbon films [3]. The resonance lines of carbon films are located near the free electron g-values. The small deviation of the resonance line position from the free electron resonance field is due to the spin-orbit coupling. In oriented carbon structures such as HOPG or carbon nanotubes the anisotropy of the spin-orbit coupling determines the anisotropy of the g-factor.

EXPERIMENTAL METHODS

Carbon thin films (100-200 nm thick) have been deposited by Low Energy Cluster Beam Deposition on silicon single crystalline surfaces as reported elsewhere [1]. The machine uses a laser ablation process (polycrystalline graphite target) to induce the carbon plasma formation with a followed supersonic expansion through the injection of a carrier gas (He) into the source. A series of Raman spectra have been obtained with a Jobin Yvon single monochromator equipped with a notch filter and a CCD detector cooled at liquid nitrogen temperature. Raman scattering has been excited through a He-Ne laser (632 nm) at low power (6 mW) to avoid heating effects. The ESR spectra were recorded by using a Bruker EMX spectrometer, operating in X-band, in the temperature range 77 K to 400 K.

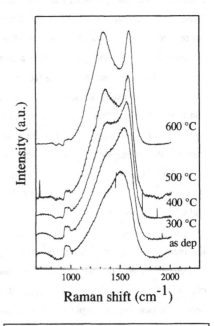

Figure 1. Some Raman spectra of as-deposited and annealed carbon nanoclusters.

EXPERIMENTAL RESULTS AND DISCUSSIONS

Thermal annealing produces deep changes in the structural and electronic configuration of amorphous carbon that are reflected in Raman spectra. Fig.1 shows some of these spectra for samples annealed for 30 min at different temperatures. As observed in Figure 2, in the domain 1000 cm^{-1} to 2000 cm^{-1}, the Raman spectrum is well described by a superposition of two Lorentzian-like lines. A graphitization process clearly starts above 350 °C as suggested by the changes induced in the 1000 cm^{-1} to 2000 cm^{-1} spectral region. This graphitization involves mainly the trigonal carbon component and is manifested through an increase of the I(D)/I(G) ratio, a blue-shift of the G line position and a sharpening of both D and G lines. In particular the blue-shift of the G line is reported in Fig.3 as a function of the annealing temperature. These changes are also accompanied by variation of other physical properties like a decreasing of the optical energy gap and the decrease of the room temperature resistivity. It is generally considered a quantitative correlation between I(D)/I(G) and the graphitic cluster correlation length (L) as follows[2]:

$$I(D)/I(G) = CL^2 \quad (1)$$

Our calculations performed on the basis of resonant data collection reported in

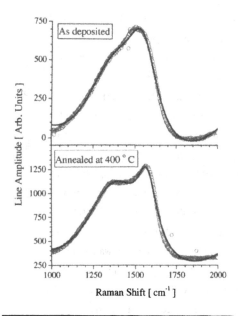

Figure 2. Simulation of Raman spectra by two overlapping Lorentzian lines

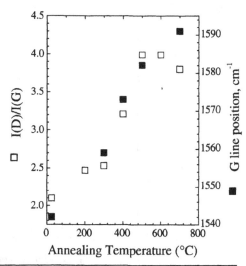

Figure 3. The effect of thermal annealing on the most important parameters of the Raman spectrum

literature have led to C=1.1 nm^{-2} for a 633 excitation wavelength.

The ESR spectra of as prepared and annealed carbon nanoclusters is a single Lorentzian line located near the free electron position. Some representative ESR spectra of as prepared and annealed carbon nanoclusters are collected in Fig. 4. It is observed that the as prepared spectrum is a convolution of Lorentzian and Gaussian lines. The thermal annealing results in broader lines with a Lorentzian shape, shifts the resonance line position towards larger g values and decrease the resonance line amplitude. A direct correlation between the G band observed in Raman spectroscopy and the ESR signal was found. Figure 5 confirms this qualitative dependence.

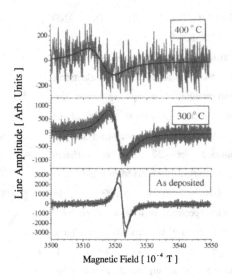

Figure 4. ESR spectra of as deposited carbon nanoclusters and some thermally annealed carbon nanoclusters.

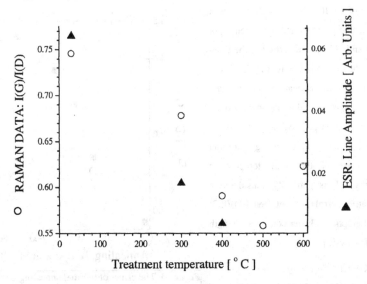

Figure 5. The correlation between ESR and Raman spectra

18

CONCLUSIONS

Raman and ESR investigations on carbon nanoclusters were reported. Two overlapping Lorentzian lines, located at about 1300 cm^{-1} and 1550 cm^{-1} were observed by Raman spectroscopy and assigned to the D and G bands of carbonaceous materials. It was proved that these lines are sensitive to the thermal annealing of the deposited carbon nanoclusters. The position of the G band was shifted towards higher wavelengths and the ratio between the amplitude of the D and G bands was increased as the annealing temperature was increased. ESR studies revealed that the as obtained carbon nanoclusters film presents a single resonance line located near the free electron position. The thermal annealing of carbon nanoclusters produces important modifications in the paramaters of the ESR line; its amplitude was decreased, the g-factor was shifted towards larger values, and the resonance linewidth is increased. These modifications rae consistent with a decrease of the concentration of paramagnetic defects.

REFERENCES

1. G. Compagnini, Mat. Res. Soc. Symp. Proc., vol.851, p.41 (2005).
2. A. C. Ferrari, J. Robertson, Phys. Rev. B 64, 075414 (2001).
3. R. C. Barklie, Diamond and Related Materials 12, 1427–1434, (2003).

Mater. Res. Soc. Symp. Proc. Vol. 887 © 2006 Materials Research Society 0887-Q01-05

Mechanical Degradation of Carbon Nanotubes: ESR Investigations

M. D. Chipara, A. Kukovecz[1], Z. Konya[1], J. M. Zaleski[2], M. Chipara[2]

PartTec Ltd, Bloomington, Indiana.
[1]Department of Applied and Environmental Chemistry, University of Szeged, Hungary
[2]Indiana University, Department of Chemistry, Bloomington, Indiana.

ABSTRACT

ESR investigations on milled multiwalled carbon nanotubes are reported. The ESR
spectra of pristine and milled carbon nanotubes consists of two resonances, a wide line
located at g > 3.0 assigned to magnetic particles (catalysts residues) and a broad line with
a peak to peak linewidth of about 10^{-2} T located at g = 2.05 and assigned to the
interaction between the conducting electrons delocalized over carbon nanotubes and
magnetic ions. It was observed that the parameters of this line (line position, linewidth,
and line amplitude) depend on the milling time. For short milling times (up to 125
minutes) both the resonance linewidth and the g-factor increase as the milling time is
increased. Longer milling times resulted in a decrease of both the g-factor and resonance
linewidth (as the milling time was increased). This behavior was assigned to the removal
of magnetic nanoparticles from carbon nanotubes. No resonance line due to the
destruction of carbon nanotubes was observed.

INTRODUCTION

The outstanding physical (Young modulus, tensile strength, thermal conductivity, and
electrical conductivity) and chemical properties of carbon nanotubes (CNTs), triggered
much theoretical and experimental research. For certain applications such as gas storage
and absorption, short opened end CNTs are required. Mechanical milling and sonication
are the most frequently utilized techniques to cut long CNTs into short ones [1-4].
Sonication in external magnetic fields has been used to purify CNTs by removing
magnetic catalyst residues [1].

TEM investigation of CNTs revealed the increase of carbon defects concentration and buckling after their sonication in solution (CH_2Cl_2) at 0 °C [3]. The sonication of CNTs produced ripple like damages (the breakage of small fraction of graphene layers) and the thinning of carbon nanotubes (by stripping the outer layers) [2].

The effect of high power sonication on freshly prepared CNTs characterized by an electron spin resonance (ESR) spectrum located at g=2.000 and characterized by a linewidth of about 30 Gauss (assigned to conducting electrons) has been reported in [2]. The resonance line of the pristine sample was reported to exhibit a reduced asymmetry [2]. The sonication of these CNTs broadened the resonance line up to about 40-50 Gauss [2]. The appearance of a new resonance line, located at g=1.991and with a linewidth of about 6 Gauss was observed [2]. The intensity of this line increases as the sonication time is increased [2]. Accordingly, this line was assigned to local defects like dangling bonds produced by sonication within the carbon nanotubes.

EXPERIMENTAL METHODS

Multiwalled carbon nanotubes were prepared by using acetylene via catalytic chemical vapor deposition at 973 K, using $CoFe/Al(OH)_3$ as catalyst The as obtained nanotubes were purified with concentrated NaOH, HCl, and $KMnO_4$-H_2SO_4 aqueous solutions, to)oremove catalysts residues and pyrolitic graphite. The energetic ball milling has been performed by using two cylindrical drums filled with multiwalled carbon nanotubes powder, placed into a vibrational mill horizontally and driven with an eccentric shaft at 50 Hz. The milling time was ranging between 0 hours and 200 hours. Further details on carbon nanotubes preparation and milling are included in [3, 4]. Electron spin resonance (ESR) investigations on milled multiwalled carbon nanotubes have been performed by using a Bruker EMX spectrometer operating in X-band (about 9 GHz).

EXPERIMENTAL RESULTS AND DISCUSSIONS

The dependence of resonance line parameters on milling time has been investigated in detail. Complex similar resonance spectra have been recorded in both pristine and milled samples. The as recorded resonance spectra of carbon nanotubes consist of two intense lines, characterized by large linewidths. We were not able to sense the narrow resonance line located at g=1.99 assigned to isolated defects (dangling like bonds) in carbon nanotubes. However, this line may be hidden in the observed broad and narrow lines. These lines were simulated by a combination of two Lorentzian-like lines [5]:

$$I(H) = I_0^{(B)} \frac{\left(\dfrac{H - H_R^{(B)}}{H_{PP}^{(B)}}\right)}{\left(1 + \left(\dfrac{H - H_R^{(B)}}{H_{PP}^{(B)}}\right)^2\right)^2} + I_0^{(W)} \frac{\left(\dfrac{H - H_R^{(W)}}{H_{PP}^{(W)}}\right)}{\left(1 + \left(\dfrac{H - H_R^{(W)}}{H_{PP}^{(W)}}\right)^2\right)^2} \tag{1}$$

Where I_0 is the resonance line amplitude, H_R is the resonance line position, and H_{PP} is the resonance linewidth. The superscripts B and W identify the broad and the wide components, respectively.

The broad line for the pristine carbon nanotubes sample is located at g=2.0425. Such a deviation from the free electron g-value ($g_{free\ electron}$=2.0023) is too large to be assigned to the spin-orbit coupling. Accordingly the broad line was assigned to the interaction between the conducting electrons localized on the conducting domains of carbon nanotubes and magnetic moments. This will shift the resonance line position to an effective g-factor, g_{eff}, defined by [6]:

$$g_{eff} = \frac{g_B \chi_B + g_M \chi_M}{\chi_B + \chi_M} \tag{2}$$

Where g_B is the g-factor of the broad line, g_M is the g-factor of the magnetic ions, χ_B is the spin susceptibility corresponding to the broad line and χ_M is the susceptibility associated to the magnetic resonance.

As it is observed from Figure 1, the g-factor associated with the broad line increases with the milling time up to about 125 minutes. This result may reflect the enhancement of the magnetic properties of catalyst residues due to the agglomeration of magnetic defects in larger particles. Above a milling time of about 125 minutes the g-factor starts to decrease, suggesting that the magnetic nanoparticles are removed from carbon nanotubes. This result is expected as both carbon nanotubes milling and sonication it is known to reduce the amount of magnetic particles attached to carbon nanotubes. The g-factor associated to the wide resonance follows the same dependence on the milling time. The sudden decrease of the g-factor of magnetic particles at large milling times may suggest that the isolated magnetic particles are converted in almost spherical particles. For such symmetry, the shape anisotropy is negligible and hence the deviation of the resonance line position from the theoretical value reflects dominantly the contribution of magnetocrystalline anisotropy (while the effect of shape anisotropy is averaged-out).

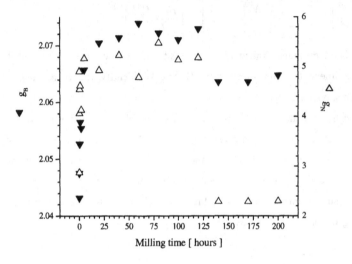

Figure 1

The wide line has a resonance line width of the order of 10^3 Gauss and originates from magnetic nanoclusters. The broad line has a linewidth of the order of 10^2 Gauss and originates from the interaction of the uncoupled electronic spins localized within the conducting domains of nanotubes with magnetic moments. As observed from Figure 2 the resonance linewidth of the broad line is enhanced as the milling time is increased up to 125 minutes. This is in agreement with the g-factor dependence on the milling time and confirms the agglomeration of magnetic nanoparticles on carbon nanotubes. In this stage the agglomeration of magnetic nanoparticles enhanced their magnetization at saturation. For longer milling times, as the magnetic nanoparticles are removed from nanotubes the resonance linewidth is decreased. The ESR data indicate that the chemical treatments did not fully remove the catalyst residues from pristine nanotubes. This is not surprising because ESR spectroscopy has outstanding capabilities in the detection of magnetic signals.

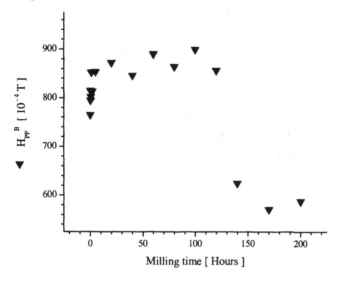

Figure 2

CONCLUSIONS

The details of the recorded resonance spectrum prevented us from an accurate estimation of the wide line parameters. The parameters of the broad resonance lines have been accurately determined. It was observed that the parameters of the broad line are affected by milling duration. That allowed us to correlate these data with the structural modifications of carbon nanotubes, as revealed by transmission electron microscopy [1, 2]. No new resonance line has been observed in milled CNTs.

REFERENCES

1. Le Thien-Nga, Klara Hernadi, Edina Ljubovic, Slaven Garaj, Laszlo Forro, Mechanical Purification of Single-Walled Carbon Nanotube Bundles from Catalytic Particles, Nano-Lett., 2(12) 1349-1352 (2002).

2. K. L. Lu, R. M. Lago, Y. K. Chen, M. L. H. Green, P. J. F. Harris, S. C. Tsang, Mechanical Damage of Nanotubes by Ultrasound, Carbon, 34(6) 814-816, 1996.

3. Z. Konya, J. Zhu, K. Niesz, D. Mehn, I. Kiricsi, Carbon, 42, 2001-2008, 2004

4. A. Kukovecz, T. Kanio, Z. Konya, I. Kiricsi, Carbon, 43, 994-1000, 2005.

5. R. C. Barklie, Diamond and Related Materials 12, 1427–1434, 2003.

6. F. Beuneu, C. l'Huillier, J.-P. Salvetat, J. -M. Bonard, L. Forro, Phys. Rev. B 15, 59, 8, 5945-5949,1999.

Mater. Res. Soc. Symp. Proc. Vol. 887 © 2006 Materials Research Society   0887-Q02-03

Effects of MeV Ions on Thermal Stability of Single Walled Carbon Nanotubes

A. R. Adhikari[1], M. B. Huang[1], C. Y. Ryu[2], P. M. Ajayan[3] and H. Bakhru[1]

[1]College of Nanoscale Science and Engineering, State University of New York, Albany, NY-12222

[2]Department of Chemistry and Chemical Biology, Rensselaer Polytechnic Institute Troy, NY-12180

[3]Department of Material Science and Engineering, Rensselaer Polytechnic Institute Troy, NY-12180

Abstract

The properties of carbon nanotubes (CNTs) are closely dependent on their structures, and therefore may be tailored by controllably introducing defects in the nanotube systems. In this work, we have investigated the effects of energetic ions (H^+ and He^+) on the thermal stability of single wall nanotubes (SWNTs) against oxidation in air. SWNTs were irradiated with MeV ions to various doses in the range 10^{13}-10^{16} cm^{-2}. Thermogravimetric analysis (TGA) was used to determine the loss of CNT masses as a result of oxidation processes. As opposed to the case of pristine SWNTs for which the temperature (T_{max}) corresponding to maximum oxidation rate was found to be ~ 495 °C, ion beam processing significantly enhanced the thermal stability of nanotubes, e.g., T_{max} increased by about 30 °C after H^+ implantation (dosage: 10^{15} cm^{-2}) and 17 °C after He^+ implantation (dosage: 10^{13} cm^{-2}). The activation energies for thermal oxidation under various conditions were also extracted from TGA data, with values ranging from 1.13 eV (for pristine SWNTs) to 1.37 eV, depending on ion doses and species. Raman spectroscopy was used to determine the characteristics of the G band (C-C stretching mode) and D band (disorder induced mode) in CNTs. The work suggests that the SWNTs modifies to more stable structure (may be cross-linked SWNTs) at small doses. Once the number of defects exceeds some critical value (depending on the type and dosage of bombarding ion) the bonding energy in CNTs weakens, leading to the reduced thermal stability of CNTs against oxidation.

Introduction

Since their discovery [1] in 1993, single-walled carbon nanotubes (SWNTs) have attracted a great attention in the scientific community even in a short span of time. May be due to unusual structure and its relation with various properties, they have become a potential candidate for a wide range of applications [2-5]. In many applications, such as high-strength composites [6] in nuclear reactors or space shuttles, CNTs are expected to expose different types of ion beams followed by interaction between them, where the stability is of primary concern since CNTs are metastable. Instead, this could modify CNTs as the interaction can produce different types of defects, like substitutional, interstitial, single or multiple vacancies, etc. [7] that plays a crucial role in their electronic properties of bulk material. Therefore, it is essential to understand the microstructural change in ion irradiated CNTs.

These days ion manipulation has become an essential process in the fabrication of Si or other compound semiconductor devices. This process introduces not only a wide range of defects in a controlled manner, but also changes their electronic and mechanical properties [8,9]. To this end, hydrogen and helium were chosen to implant into CNTs as they are smaller than a carbon atom, and subsequently introduce small number of defects. It is believed that a large number of defects could bring structural modification. The main objective of this work is to present the oxidation behavior of non-implanted and different ions (Hydrogen, Helium) implanted SWNTs. Raman spectroscopy was also used to investigate the microstructural change by monitoring the structural lattice damage, via changes in phonon mode spectra resulting from irradiation.

Experiment

Purified Single walled carbon nanotubes (BuckyPearls) used for the present study were produced by high pressure CO disproportionation (HiPco) at CNI, Houston, USA. BuckyPearls were crushed into powder using mortar and pestle in an Acetone and made suspension using ultrasonic agitation for about 30 minutes. Finally, samples were prepared by suspension cast method on a cleaned silicon wafer and allowing them to dry in an open air. The implantations were then performed with ions (Hydrogen or Helium) with doses in the range 10^{13}-10^{16} cm^{-2} at ambient temperature using 4 MV dynamitron accelerator. Different ions of different energies were used to produce wide range of damage to a depth of ~5.35 μm based on Monte Carlo simulation. Table 1 gives the details of the ion irradiation condition employed.

Thermo Gravimetric Analyzer (Mettler-Toledo TGA/SDTA851e) was used to study the thermal oxidation behavior in air between temperatures 200-700 °C with the heating ramping rate of 9 °C min^{-1}. Raman scattering measurement was performed using a Renishaw Raman microscope, where the polarized Ar ion laser beam of excitation wavelength of 514.5 nm operating at 24 mW power was used. The back-scattered radiation was collected and analyzed using a double-grating monocromator, where the phonon spectrums were monitored in the range of 180 to 3300 cm^{-1}.

Table 1. Ion irradiation conditions used in the experiment where the values were calculated using the TRIM code.

Ion	Energy (MeV)	Range (μm)	$(dE/dx)_e$ (eV/A)	$(dE/dx)_n$ (10^{-3} eV/A)	Carbon vacancies/A/ion
^{1}H	0.4	5.36	5.48	4.51	45×10^{-5}
^{4}He	1.2	5.34	23.3	24.3	0.007

Results and discussion:

The TGA thermogram provides the quantitative information on the weight change at different bonding temperatures. TGA analysis was made for all unimplanted and 0.4 MeV Hydrogen implanted SWNTs (H-SWNTs) under the flow of air at 50 ml/cm^3. Thermograms of all the implanted samples showed similar nature with the unimplanted SWNTs, a single step degradation (Fig. 1). From the observation of % weight loss at different temperatures, similar oxidation behavior was observed until 350 °C, as all samples implanted with different doses are equally resistive up to this temperature, although there can form different types of defects such as single or multiple vacancy, substitutional defects, etc. These defects, along with as grown defects, can be the source of degradation likely at higher temperature. They start to show changes as the temperature increases beyond 350 °C. By comparing to the % mass left, H-SWNTs with a dosage of 1×10^{15} cm^{-2} showed the least oxidation rate, however after a dosage of 1×10^{15} cm^{-2} oxidation began to increase due to some disorder, possibly the scission of highly strained crosslink to the system.

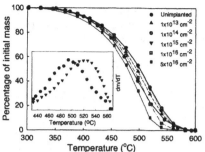

Figure 1. TGA thermogram for unimplanted and Hydrogen implanted SWNTs. The inset shows the TGA derivative curve for unimplanted and Hydrogen implanted (dosage: 10^{15} cm^{-2}) SWNTs. Note that the thermogram does not include residue (catalyst) of the pyrolysis.

Activation energy is one of the useful parameter for studying the thermal stability of the material. Although, numerous methods have been proposed, Briodo relation [10] derived under a dynamical condition was used to extract the activation energy in the present investigation. A firm constant value of activation energy was found until the dosage of 1×10^{15} cm^{-2} and began to increase as shown in Fig. 2a. We prepared another set of SWNTs samples implanted with 1.2 MeV Helium (He-SWNTs) for all doses from 10^{13} cm^{-2} to 10^{16} cm^{-2}. An analogous interaction behavior was observed. However, the activation energy begins to increase even at the smallest dose and at higher doses, the activation energy approaches to the value corresponding to that of 2-D graphite [11,12] - as the graphite is the most stable form of carbon. For Helium, the maximum stability is observed at a dosage of 1×10^{13} cm^{-2} unlike the case of Hydrogen, where the stability is maximum at 1×10^{15} cm^{-2}. This decrease of dose corresponding to the maximum stability could be due to their higher interaction cross section compared to Hydrogen ions, since 1.2 MeV Helium can induce a damage of 15.5 vacancies/A/ion compared to 0.4 MeV Hydrogen, thus expecting smaller Helium doses compared to Hydrogen to create same level of defects.

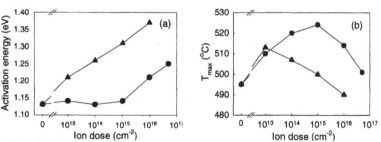

Figure 2. Plot of activation energy (a) and T_{max} (b) for pristine and implanted SWNTs with Hydrogen (●) and Helium ions (▲).

The shift of T_{max} to higher temperature means higher thermal stability. For a stability comparison, T_{max} was plotted for Hydrogen and Helium implanted samples as obtained from the TGA plot (Fig. 2b). The plot illustrates the stability enhancement in both cases at lower doses. The stability for Hydrogen implanted samples was found higher compared to Helium implanted samples i.e,

T_{max} shifted by about 30 °C for H-SWNTs (dosage: 1×10^{15} cm^{-2}) and about 17 °C for He-SWNTs (dosage: 1×10^{13} cm^{-2}). The exact reason of T_{max} shift to higher temperature after Hydrogen and Helium implantation is not clearly known at this time.

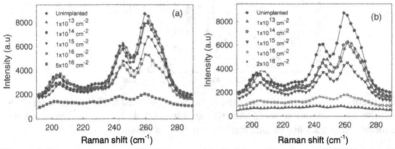

Figure 3. Room temperature RBM spectra of pristine and implanted SWNTs with Hydrogen (a) and Helium (b).

One of the first orders Raman spectra is the radial breathing mode [13] (RBM) in the range of 170–280 cm^{-1}. The characteristic feature of RBM is its frequency dependence on the nanotube diameter and provides the quantitative information of SWNTs in regard to its structure. Fig. 3 demonstrates the radial mode of Raman spectrum of unirradiated and irradiated SWNTs. SWNTs showed complete resistant for all doses of Hydrogen (Fig. 3a), although the stability decreases after a dosage of 1×10^{15} cm^{-2}. In case of Helium (Fig. 3b), the radial mode is seen just surviving after a dosage of 1×10^{16} cm^{-2}. Here, it is expected that the SWNTs nearly lose its intrinsic cylindrical structure at this highest Helium dose.

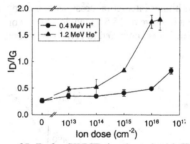

Figure 4. Schematic variation of I_D/I_G for SWNTs implanted with Hydrogen and Helium for different doses from 10^{13} to 10^{16} cm^{-2}.

Another two prominent features of the 1st order scattering are disorder induced D-band [13], due to the finite particle size effect or lattice distortion and tangential vibrational mode of graphite, G-band [13]. With the irradiation, D-band intensity is seen increasing following decrease in G-band intensity after certain dose depending on types of ions. Fig. 4 showed that for Hydrogen and Helium, I_D/I_G values increases with the dose implies that the number of defects is increasing. The I_D/I_G value varies inversely with the crystallite size (L_a) of the graphite sample using Tuinstra-Koenig relation [14] (TK) for different stages of disorder. Using TK relation, the I_D/I_G value is expected to increase to a maximum, when the material amorphise with the smallest aromatic clusters. The I_D/I_G value (for Hydrogen and Helium) found increasing with the dose, conforming that the infinitesimal symmetrical cylinder is disintegrating, such that the L_a is decreasing in both cases.

To better understand the effect of the ion doses on the structural modification and the bonding configurations of carbon atom resulting from ion irradiation, we monitored the variation of G and D peak position for various ion implanted samples as shown in Fig. 5. Fig. 5a (for H-SWNTs) shows a fairly constant value within the spectral resolution at lower doses ($\leq 1\times10^{15}$ cm^{-2}) - showing the bond rigidity for these ion doses, although their interaction can form different types of defects that lower the crystal symmetry. However, a clear signal of downshift at higher doping level is observed in both the G and D-band position. A similar behavior was observed from Helium implanted samples (Fig. 5b).

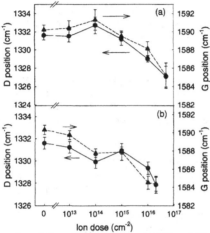

Figure 5. Variation of G peak and D peak position for (a) H$^+$ and (b) He$^+$ implanted SWNTs for various doses from 10^{13} to 10^{16} cm^{-2}.

There have been reported some degree of stability enhancement in the past using different approaches [15-17]. These experiments showed that stability enhancement could be either due to reduction of defect sites or the reduction of iron content or both. However, our experiment was carried at a temperature below 600 °C, unlike the higher temperature in the studies mentioned above [15-17]. So the stability enhancement in our experiment is due to some factor other than annealing defects, removal of catalyst particles or the interaction with the high-Z element. For both ions, thermal stability is increased at low doses. The smaller I_D/I_G value at these doses implies that the stability is prominent only at low defect density. According to theoretical modeling [8,18], both the vacancy and carbon recoil that are the most prevalent in irradiation are more mobile than the coplanar counterpart. It was shown that the mending of vacancies create a pentagon with adjacent carbon atom containing dangling bond displaced outwards and this is believed to take part in cross-linking when divacancies approaches. Similarly, carbon recoils are also found very sensitive and make a four fold coordinated bridge even at room temperature. Thus, the stability enhancement at small doses in our case is expected as an intriguing phenomenon with relatively low defects concentration leading to polymerization through inter-tube sp^3 bonding resulting in cross-linking between CNTs or CNT and nanoparticles.

Conclusion:

SWNTs oxidation has been investigated with respect to thermal stability in the temperature range of 200-700 °C based on remaining weight at various decomposition temperatures, T_{max} and E_a. This TGA investigation showed characteristic behaviors for the implanted impurities, where ion doses seem to have the most significant effect on thermal stability, and provides the convincing evidence for irradiation-mediated improvements of the thermal stability of SWNT bundles at lower doses of Hydrogen and Helium. The results also showed that stability is maximum after the Hydrogen dosage of 1×10^{15} cm^{-2} and after the Helium dosage of 1×10^{13} cm^{-2}. Although, it is difficult to propose a clear conclusion in regard to stability enhancement, we nonetheless conclude it to be due to cross-linking between CNTs or CNT and impurities present in the sample and it needs further study to use this behavior on various technological applications.

[1] S. Iijima, T. Ichihashi, *Nature* **363**, 603 (1993).
[2] P. G. Collins, A. Zettl, H. Bando, A. Thess and R. E. Smalley, *Science* **278**, 100 (1997).
[3] E. W. Wong, P. E. Sheehan and C. M. Liber, *Science* **277**, 1971 (1997).
[4] P. Chen, X. Wu, J. Lin and K. Tan, *Science* **285**, 91 (1999).
[5] M. S. Dresselhaus, G. Dresselhaus, P. Avouris, *Carbon Nanotubes*, Vol. **80** (Springer-Verlag, Heidelgurg, 2001).
[6] B. Lahr, J. Sandler, *Kunstsoffe* **90**, 94 (2000).
[7] A. V. Krashminnikov and K. Nordlund, *J. Vac. Sci. Technol.* B **20**, 728 (2002).
[8] J. Kotakoski, A. V. Krasheninnikov, Y. Ma, A. S. Foster, *Phys. Rev. B* **71**, 205408 (2005).
[9] M. Bockrath, W. Liang, D. Bozovic, J. H. Hafner, C. M. Liber, M. Tinkham and H. Park, *Science* **291**, 283 (2001).
[10] A. Broido, *J. Polym. Sci.* **7**, 1762 (1969).
[11] X. Chu and L. D. Schmidt, *Carbon* **29**, 1251 (1991).
[12] D. W. Mckee and C. L. Spiro, *Carbon* **23**, 437 (1985).
[13] Dresselhaus M S, Dresselhaus G, Eklund P C 1996 *Science of Fullerenes and Carbon Nanotubes*, Academic Press, San Diego
[14] F. Tuinstra and J. L. Koening, *J. Chem. Phys.* **53**, 1126 (1970).
[15] E. Salonen, A.V. Krasheninnikov and K. Nordlund, *Nucl. Instr. and Meth.* B **193**, 603 (2002).
[16] M. J. Lopez, A. Rubio, J. A. Alonso, S. Lefrant, K. Metenier and S. Bonnamy, *Phys. Rev. Lett.* **89**, 255501 (2002).
[17] D. Bom, R. Andrews, D. Jacques, J. Anthony, B. Chen, M. S. Meier and J. P. Seleque, *Nano Lett.* **6**, 615 (2002).
[18] A. A. El-Barbary, R. H. Telling, C. P. Ewels, M. I. Heggie and P. R. Briddon, *Phys. Rev. B* **68**, 144107 (2003).

Mater. Res. Soc. Symp. Proc. Vol. 887 © 2006 Materials Research Society   0887-Q06-03

Investigations of *dc* electrical properties in electron-beam modified carbon nanotube films: single- and multiwalled

S. Gupta[1,*], N. D. Smith[1], R. J. Patel[2], and R. E. Giedd[2]
[1] Department of Physics, Astronomy, and Materials Science
[2] Center for Applied Science and Engineering, Missouri State University, Springfield, MO 65897 (USA).

ABSTRACT

Carbon nanotubes (CNTs) in the family of nanostructured carbon materials are of great interest because of several unique physical properties. For space applications, it needs to be shown that CNTs are physically stable and structurally unaltered when subjected to irradiation becomes indispensable. The CNT films were grown by microwave plasma-assisted chemical vapor deposition (MWCVD) technique using Fe as catalyst. Synthesis of both single- and multiwalled CNTs (SW and MW, respectively) were achieved by varying the thickness of the Fe catalyst layer. To investigate the influence of electron-beam irradiation, CNTs were subjected to low and/or medium energy electron-beam irradiation continuously for a few minutes to several hours. The CNT films prior to and post-irradiation were assessed in terms of their microscopic structure and physical properties to establish *property-structure* correlations. The characterization tools used to establish such correlations include scanning electron microscopy (SEM), high resolution transmission electron microscopy (HRTEM), Raman spectroscopy (RS), and current *versus* voltage (*I-V*) measuring resistance (*two-probe*) and conductivity (*four-probe*) properties. Dramatic improvement in the *I-V* properties for single-walled (from semiconducting to quasi-metallic) and relatively small but systematic behavior for multi-walled (from metallic to more metallic) with increasing irradiation hours is discussed in terms of critical role of defects. Alternatively, contact resistance of single-walled nanotubes decreased by two orders of magnitude on prolonged E-beam exposures. Moreover, these findings provided onset of saturation and damage/degradation in terms of both the electron beam energy and exposure times. Furthermore, these studies apparently brought out a contrasting comparison between mixed semiconducting/metallic (*single-walled*) and metallic (*multiwalled*) nanotubes in terms of their structural modifications due to electron-beam irradiation.

INTRODUCTION

Advanced nanostructured carbon-based materials continue to attract attention because of their unique combination of structural and physical properties and therefore offering multifunctionality. CNTs, in particular, exhibit exceptional structural (high aspect ratio of ~ 1000), physical (electrical, electronic, mechanical), and chemical properties, which have resulted in their continued investigations since they were discovered [1, 2]. One of the exciting behaviors of SW is their dual electronic character (semiconducting and metallic) which is based on their chirality. In other words, SW can be used for semiconducting devices as well as conducting nano-wires if sought for [2]. In either case *i.e.* semiconducting or metallic, it is possible that CNTs could be used as an electronic device component which may be subjected to unavoidable radiation, aboard a spacecraft bound for outer space where the atmosphere no longer provides protection against space radiation, for instance. Under these conditions, it would be useful to

---

* To whom the correspondence should be addressed, E-mail: SGupta@MissouriState.edu.

know how the CNTs would behave if subjected to such an environment. In the present study, the space radiation environment is terrestrially simulated using a beam of electrons generated by thermionic emission from a tungsten filament and the E-beam energy is 30 keV and 100-200 keV from SEM and TEM electron-gun, respectively. Although the former energy is relatively low compared to most space radiations such as cosmic rays (on the order of MeV), and unlike previous experiments [3,4,5,6] where high energy radiation was used, the CNTs degrade rapidly in the latter case. The relatively low electron beam energy studies allow the relation between the E-beam exposure time and the extent of damage/degradation, since low energy E-beam causes the degradation process to occur rather slowly [7, 8]. The systematic approach of irradiating the nanotubes (both the SW and MW) with E-beam continuously for 2.5, 5.5, 8, and 15 hours and characterizing their nano/microscopic structure and electrical properties (both pristine and irradiated) permitted us to draw conclusions in terms of structural transformation/alteration and defects associated with them.

Experimental studies show MW tends to be relatively more robust than those of SW. This is because increased exposure on an occasionally found individual bundle of SW tend to graphitize, pinch, and cross-link similar to polymers forming intra-molecular junction (IMJ) within the area of electron beam focus, which occurs possibly through aggregates of amorphous carbon [7, 8], revealed through SEM and TEM micrographs. It is also suggestive that knock-on collision may not be the primary cause of structural degradation, rather but a local gradual re-organization occurs. For all of the measurements, resistance values lie in a range of 3 - 10 kilo-ohms at room temperature. The results also indicate that MW tend to reach a state of damage saturation, suggesting the possibility of developing radiation shields from MW for short-term space missions and electromagnetic interference shielding (EMI) as well. Conversely, SW are not one of the promising and reliable candidates in radiation environment for long-term space applications such as electronic devices and/or possibly radiation hard programmable logic circuits [9] without fail, but perhaps implemented as radiation sensors [8].

EXPERIMENTAL PROCEDURES

The CNT films used in this experiment were grown by microwave plasma-assisted chemical vapor deposition (MWCVD) employing a mixture of acetylene and ammonia as precursor gases [10]. They were grown on silicon (100) substrates deposited with iron layer as catalyst to promote growth of nanotubes. Different thicknesses of iron were deposited onto silicon using electron beam evaporator. Thinner iron (~ 0.5 nm) promoted the growth of SW, while MW nanotubes were grown with thicker iron (~ 80nm) layer. Once synthesized, the CNT films were subjected to a constant beam of electrons with 30 keV (or energy fluence of 30 GeV/cm^2) for 2.5, 5.5, 8, and 15 hours continuously. The E-beam was generated by thermionic emission from a tungsten filament of a JEOL JSM 6360LV, a low vacuum scanning electron microscope and 100-200 keV from TEM (Model JSM 2000x) electron gun.

Complementary analytical tools such as scanning electron microscopy (SEM), high resolution transmission electron microscopy (HRTEM), and Raman spectroscopy (RS) were employed to probe radiation-induced structural modifications. SEM micrographs were imaged with secondary electron detector and beam energy of 5 kV. The same scanning electron microscope used to irradiate the samples (alternatively, the experiments were conducted *in situ*). In the mean time, the HRTEM micrographs were obtained on a Philips EM430 Super Twin with beam energy between 200 and 300 kV. Resonance Raman spectroscopy measurements were made using an Ar$^+$ laser excitation source with a wavelength of 514.5 nm ($E_L = 2.41$ eV) and incident power ~ 5 mW to avoid thermal degradation, an ISA-JY Triax 320 spectrometer, and an ISA-JY Spectrum-One CCD 3000 charge coupled device as the detector. All of the Raman spectra were measured in backscattered configuration and fitted using Jandel Scientific Peakfit software (v. 4.0) based on Marquardt-Levenberg method [11]. The *dc* electrical properties were determined from *I-V* measurements using both the two-probe and four-probe methods. The two-probe setup consisted of a Keithley 230 programmable voltage source in series with a Keithley 6517a electrometer (ammeter). Two finely tipped probes (~ 5 µm at tip) placed approximately 5 mm apart, touch the surface of the sample and an electric field is applied, which in turn causes a current flow measured by the digital ammeter. The four-probe measurements were carried out using a Keithley 240 voltage and ammeter source. For the electrical resistivity, a probe station (Model Signatone) composed of a Keithley 224 programmable current source unit and a Keithley digital voltmeter 196 DMM were used. The electrical conductivity was measured using the four-probe in Van *der* Pauw metal-CNT–Fe/SiO$_2$/Si configuration. These measurements permitted us to determine the electrical conductivity ($\sigma = 1/\rho$) [12].

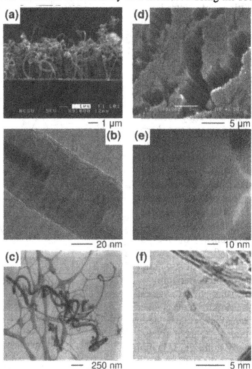

RESULTS AND DISCUSSION

The SEM micrograph of pristine MW (Fig. 1a) reveals a complicated surface morphology with an areal density of ~ 10^8/cm^2, whereas the SW (Fig. 1d) have a much higher areal density in the range from 10^{13} to 10^{14}/cm^2. Also as seen from Figure 1, the CNT films are both vertically aligned; however, SW possess carpet-like surface morphology except where there is an occasional gap. Quantitatively, the TEM images of pristine nanotubes show diameters to be ~ 60 - 80 nm with inter-tube wall spacing ~ 0.32 nm for the MW. In contrast, the diameter of the SW ranged from 0.97 - 1.24 nm. These values agree with those determined from the radial breathing mode in the Raman spectra [13]. Qualitatively, the TEM micrographs of pristine MW exhibit compartment-like (Fig. 1b) and hollow structures (Fig. 1e) in the case of SW. Each displays wall-rippling, and in either case is attributed to short term E-beam exposure thus it is an imaging artifact. For longer exposure

Figure 1 Typical examples of pristine NTs versus E-beam modified. SEM and TEM Micrographs portray pristine vertically-aligned (a, b) MW and (d, e) SW nanotubes. Corresponding TEM images of E-beam structural modifications displaying wall rippling, curling, and T-junction are also shown as an example for (c) MW and (f) SW nanotubes, respectively.

times ($\geq$ 8 hours), the SW start to crosslink forming T-junctions (Fig. 1f), and MW forms nano-helixes (Fig. 1c), a result of both the growth and irradiation (ca. Fig. 1a with 1c) [8]. These studies allowed favorable electrical contacts through local 'soldering'.

Through the selection of excitation wavelength λ_L = 514.5 nm (E_L = 2.41 eV) for Raman spectroscopy, we preferentially probe semiconducting tubes based on the zone-folding scheme for SW unlike MW, which are invariably metallic [14,15]. Assignment of the vibrational features in the Raman spectra is carried out following Ref. 14. Two of the most pronounced features of interest in the high frequency first-order Raman spectra shown in Figure 2 are D and G bands occurring ~ 1340 cm^{-1} and 1580 cm^{-1}, respectively. The latter mode results from the splitting of intra-layer stretching mode in graphite possessing E_{2g2} symmetry and due to curvature induced re-hybridization, usually located at ~ 1560 cm^{-1} (T_2 mode) and ~ 1593 cm^{-1} (T_3 mode). They seem to be quite narrow and sharp indicating the high uniformity of the pristine SW, low level of impurities, and primarily semicon-ducting in nature [13]. The ratio of the intensity of the D band to the intensity of the G band (I_D/I_G) is a measure of the degree of structural disorder/imperfection, such that lower values correspond to greater order with much less defects. Quantitatively, for MW the I_D/I_G ratio varies from 0.82 to 0.85 as a function of E-beam

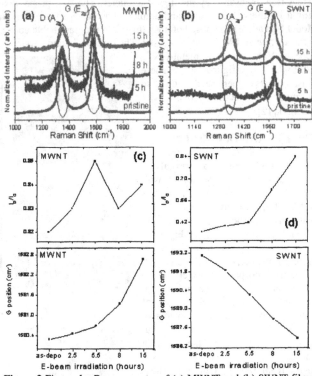

Figure 2 First-order Raman spectra of (a) MWNT and (b) SWNT films prior to and post e-beam irradiation. The spectra are shifted vertically for clarity. The I_D/I_G ratio and the G band positions as a function of E-beam exposure times are plotted for the (c) MWNTs and (d) SWNTs.

irradiation exposure time (see Fig. 2c), which is quite marginal. Whereas the I_D/I_G ratio for the SW varies from about 0.4 to 0.8 (see Fig. 2d) which is a substantial change. In addition, the position of the G band or tangential displacement mode blue-shifted 1580.3 $\rightarrow$ 1582.6 cm^{-1} for MW (see Fig. 2c) and red shifted 1593.0 $\rightarrow$ 1586.2 cm^{-1} (see Fig. 2d) for SW nanotubes. The former result points at the development of a small compressive stress due to E-beam irradiation, while the latter suggests a change from semiconducting to metallic in behavior, or graphite-like, which agrees with the results observed in *I-V* measurements. Notice the shape of the G band, which is asymmetric for metallic nanotubes - similar to Fano lineshape [15] (see Fig. 2 post-irradiation). This response is expected for single-wall nanotubes because the collapse leads to graphitization (alternatively, radiation-induced amorphization) and possibly transforms into a multi-wall nanotube *i.e.* forming larger diameter tubes.

The *dc* electrical properties of the CNT films were investigated using both the two- and four-probe *I-V* measurements. The two-probe method involves a voltage source and an ammeter connected in series as shown in Fig. 3a. *I-V* characteristics for the MW nanotube films (see Fig. 3b) prior to and post E-beam irradiation show an increase in slope with increasing irradiation time. This suggests a decrease in resistance by one order of magnitude. Notice that for exposure times of ≤ 8 hours of irradiation, the *I-V* curves for SW nanotube films (see Fig. 3c) exhibit the characteristic S-curve usually associated with semiconductors. While for 15 hours of irradiation, *I-V* curves exhibit straight line behavior associated with materials such as metals, supporting the '*claim*' of irradiation induced quasi-metallic SWNT trend.

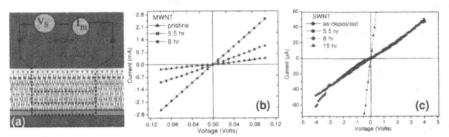

Figure 3 (a) Schematic of two-probe setup used to measure *I-V* curves for (b) MW and (c) SW prior to and post-irradiation times of 5.5, 8, and 15 hours showing the variation in contact resistance.

Figures 4b and 4c display the results of four-probe *I-V* measurements yielding electrical resistivity (alternatively, conductivity). The logarithmic plot of the *dc* conductivity *versus* the E-beam irradiation time reveals the increase in conductivity by two orders of magnitude for both the SW and MW nanotubes. Notice that the MW have a much greater conductivity as anticipated since they are invariably metallic in nature, whereas the SW films are a mixture of semiconducting and metallic nanotubes. The conductivity of the MW seem to be unaffected for irradiation times of < 5.5 hours and the SW experience an abrupt change in conductivity for exposure times ≥ 8 hours. One may speculate that the MW is reaching a state of saturation /degradation (conversely, radiation resistance) for ~ 15 hours of irradiation as demonstrated by leveling off of the *dc* conductivity. Whereas the SW experience a sharp rise in *dc* conductivity between 8 and 15 hours of irradiation providing us a simultaneous indication for both the electron energy and exposure time thresholds. Furthermore, to investigate the usefulness of the present technique for making contacts, as mentioned above (see Fig. 1) in a nanotube network, measuring the electrical properties of such a coherent junction between SW and MW would be a challenge for future studies. These results are unprecedented and more importantly, demonstrate an interesting and contrasting comparison between the single-walled (primarily semiconducting) and multi-walled (purely metallic) nanotubes.

CONCLUSIONS

In summary, the influence of electron beam irradiation on the structural and electrical properties of SW and MW carbon nanotube films is investigated. SW nanotubes experienced by far the greatest modifications which led to property alterations in addition to nano-manufacturing. The changes in SW films from semiconducting to quasi-metallic discard their candidacy for long-term space mission applications, unless they could be calibrated to measure radiation levels in space. From Raman spectroscopy, it is suggestive that a local gradual reorganization occurs *i.e.* sp^3, $sp^2 \Leftrightarrow sp^{2+\delta}$. In addition, there may be formation of pentagon and heptagon pairs (or Stone-

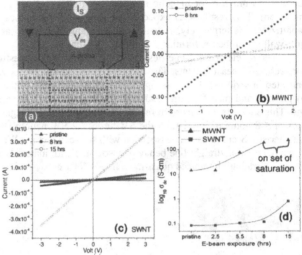

Figure 4 (a) Schematic of four-probe used to measure *I-V* curves for (b) MW and (c) SW prior to and post-irradiation times of 2.5, 5.5, 8, and 15 hours, from which the (d) *dc* electrical conductivity is determined.

Wales defects) [15] serving as a source of formation of other nanocarbons (nano- helixes, nano-T-junctions, and so forth). MW nanotubes seem to be relatively more robust and exhibit onset of saturation for prolonged exposures. Consequently, multi-walled nanotubes are proposed as radiation shields for short-term space missions and electromagnetic interference shields. These findings indeed provided a contrasting comparison between single-walled (most likely, *semiconducting*) and multi-walled (*metallic*) carbon nanotubes. It would be useful and interesting to explore their composites with polymers as reinforced agents for the above mentioned applications.

ACKNOWLEDGEMENTS

The corresponding author wishes to acknowledge her collaborators including Drs. R. J. Nemanich and Y. Y. Wang (NC State University, Raleigh-Physics Department) for sharing some of the carbon nanotube samples used in the present study. We also thank the facilities housed in the Center for Applied Science and Engineering (CASE) directed by one of the authors (REG) and ONR Grant No. N00014-03-1-0893 for financial support.

REFERENCES

1. S. Iijima, Nature 354, **56** (1991); R. H.Baughman, A. A. Zakhidov, and W. A. de Heer, Science **297**, 787 (2002).
2. M. S. Dreselhaus, G. Dresselhaus, and P. C. Eklund, in *Science of Fullerenes and Carbon Nanotubes*, Academic Press Inc. San Diego, Ch. 19 (1996).

3. J. Li and F. Banhart, Nano Lett. **4**, 1143 (2004); F. Banhart, Nano Lett. **1**, 329 (2001).

4. V. H. Crespi, N. G. Chopra, M. L. Cohen, A. Zettl, and S. G. Louie, Phys. Rev. B **54**, 5927 (1996); N. G. Chopra, L. X. Benedict, V. H. Crespi, M. L. Cohen, S. G. Louie, and A. Zettl, Nature **377**, 135 (1995); L. X. Benedict, V. H. Crespi, N. G. Chopra, M. L. Cohen, S. G. Louie, and A. Zettl (unpublished).

5. D. Ugarte, Nature **359**, 707 (1992).

6. B. W. Smith and D. E. Luzzi, J. Appl. Phys. **90**, 3509 (2001).

7. S. Gupta, R. J. Patel, N. D. Smith, Mater. Res. Soc. Symp. Proc. **851**, NN6.3 (2004).

8. S. Gupta, R. J. Patel, N. D. Smith, R. E. Giedd, and Y. Y. Wang, J. Appl. Phys. (2005) (*in press*).

9. B. M. Segal, Nantero Inc. Woburn, MA (www.nanotero.com).

10. Y. Y. Wang, S. Gupta, R. J. Nemanich, Z. J. Liu, and L. C. Qin, J. Appl. Phys. **98**, 014312 (2005) and references therein; S. Gupta, B. L. Weiss, B. R. Weiner, L. Pilione, A. Badzian, and G. Morell, J. Appl. Phys. **92**, 3311 (2002) and references therein.

11. D. W. Marquardt, J. Soc. Indis. Appl. Math. **11**, 431 (1963).

12. S. Gupta, N. D. Smith, R. J. Patel, and R. E. Giedd, J. Mater. Res. (2005) (submitted).

13. Y. Y. Wang, S. Gupta, and R. J. Nemanich, Appl. Phys. Lett. **85**, 2610 (2004).

14. A. M. Rao, E. Richter, S. Bandow, B. Chase, P. C. Eklund, K. A. Williams, S. Fang, K. R. Subbaswamy, M. Menon, A. Thess, and R. E. Smalley, Science, **275**, 187 (1997).

15. M. S. Dresselhaus and P. C. Eklund, Adv. Phys. **49**, 705 (2000).

Mater. Res. Soc. Symp. Proc. Vol. 887 © 2006 Materials Research Society 0887-Q03-04

Electron Spin Resonance on Carbon Nanofibers

Mircea Chipara[1], Robert H. Hauge[2], Hua Fan[2], Richard Booker[2], Haiqing Peng[2], Wen Fang Hwang[2], J. M. Zaleski[1], Richard Smalley[2]

[1]*Indiana University, Bloomington, IN 47405*
[2]*Chemistry Department, University of Rice, Houston, TX*

INTRODUCTION

The discovery of carbon nanotubes (CNT) and the subsequent evaluation of their outstanding properties (huge Young modulus, high electric conductivity, and large thermal conductivity) incited the research of composite materials obtained by dispersing CNT in different polymeric matrices. An increase of the Young modulus and tensile strength of the polymeric matrix, due to the dispersion of low amounts of CNT (typically below 10%) was reported [1-3]. Extremely low percolation thresholds (even smaller than 1 % wt CNT in polymers) have been reported in various composites based on polymeric matrices [1, 3]. It was speculated [1-3, 8, 9] that the high aspect ration of nanotubes would trigger anisotropic properties in polymer-carbon nanotubes materials that would allow a further enhancement of their mechanical, electrical, and thermal properties. To obtain a composite with excellent properties is required to have uniform dispersion of CNT (for isotropic composites) or controlled orientation of CNT (anisotropic composites). The agglomeration of CNTs makes difficult their dispersion in polymer solutions or melts [3]. Carbon nanofibers (CNF) is a collection of SWNTs oriented parallel dispersed in a polymeric matrix. This is a cheaper approach [8] to anisotropic composites, lightweight composites, high-strength composites, and electrical (thermal) conducting composites than the standard dispersion of randomly oriented and agglomerated SWNTs.

The as prepared (crude) CNF shows weak metallic properties resulting from the p doping (presumably due to bisulfate ions [8, 9]). Electron spin resonance (ESR) consists in the resonant absorption of energy by an ensemble of uncoupled electronic spins located in an external magnetic field from the magnetic component of an electromagnetic field. The resonance condition equals the Zeeman splitting (due to the external magnetic field) with the frequency of the electromagnetic field (in microwaves the typical resonance field is about 0.35 T-for a free uncoupled electronic spin). Hence, ESR is a versatile technique that allows the study of conducting materials. Some authors [6] use a different terminology defining by conduction electron spin resonance the ESR spectroscopy of conducting materials.

EXPERIMENTAL METHODS

Crude CNF obtained from purified HiPco SWNT [8, 9] were investigated by ESR spectroscopy using a Bruker EMX spectrometer operating in X-band (9 GHz). The angular dependence of the resonance line has been investigated. The preferential orientation of SWNTs is confirmed by the angular dependence of the g-factor and of the resonance line width.

EXPERIMENTAL RESULTS AND DISCUSSIONS

A typical ESR spectrum of crude CNF is shown in Fig. 1. It is observed that the resonance spectrum consists of a broad, intense, and asymmetric line. A faint and broad resonance line due to magnetic impurities is observed at low magnetic fields. The resonance spectrum is dominated by an intense asymmetric line, which be assigned either to the convolution of several symmetric resonance lines or to a Dysonian resonance line. The Dysonian resonance line is a typical asymmetrical resonance line observed in conducting systems. The distortion of the symmetrical resonance line (reported in insulators) reflects the damping of the electromagnetic field within the conduction domain. For carbon nanotubes the typical skin depth is of the order of about 100 μm. Hence, if the nanotubes are oriented along the microwave field the skin depth is comparable with the conducting domain length and the Kittel's theory of the anomalous skin [6] effect may be used to simulate the resonance line shape;

$$
y = C \left\{
\begin{array}{l}
\left(1 - a^2\right) \dfrac{\left[2 + \left(1 + x^2\right)^{1/2} \left[\left(1 + x^2\right)^{1/2} - 1\right]\right]^{1/2}}{\left(1 + x^2\right)^{3/2}} \, sign \, (x) + \\
+ a \dfrac{\left[\left(1 + x^2\right)^{1/2} - 2\left[\left(1 + x^2\right)^{1/2} + 1\right]\right]}{\left(1 + x^2\right)^{3/2}}
\end{array}
\right\}
\tag{1}
$$

where x=K(H-H$_R$), C and K are constants, H is the external magnetic field, and H$_R$ is the resonance field. The parameter "a" controls the resonance line shape. It is equal to the ratio between the real and the imaginary components of the impedance of the sample. Dysonian like resonance lines were reported [7] in graphite.

Actually the recorded spectrum is more complex and cannot be accurately fitted by the above expression. The resonance line shape of CNF was accurately described by a convolution of two Dysonian lines (and an additional Lorentzian line due to magnetic impurities). As may be observed from Figure 1A, an excellent agreement between the recorded resonance spectrum and the simulated one was obtained.

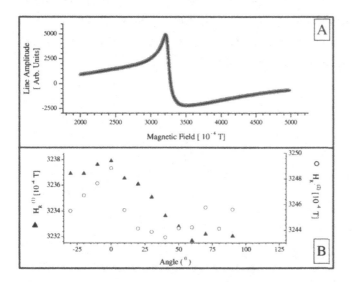

Figure 1. (A). The ESR spectra on CNF. The gray line is the recorded spectrum and the black one the simulated one. (B). The angular dependence of the central resonance lines

A characteristics of ESR spectra of conducting systems (including graphite, CNT, and CNF) is the shift of the g-factor from the theoretical value $g_0=2.0023$ due to the spin-orbit coupling [7]. In highly oriented pyrolytic graphite (HOPG) the parallel component of the g-factor (the external magnetic field is along the c axis of HOPG) is larger than the perpendicular one and this anisotropy increases as the temperature is decreased [4, 7].

The preferential dependence of carbon nanotubes inside nanofibers is supported by the angular dependence of the resonance field, shown in Fig. 1B for both lines that contributes to the central part of the resonance line. Figure 2 shows the angular dependence of the resonance line width, H_{PP}, of these 2 components. The broadest line has high shape anisotropy and originates from electrons delocalized over domains with a high electrical conductivity. Tentatively this line is assigned to electrons delocalized over the conducting domains of CNT. The origin of the second line is not yet clear. It is possible to assign this line to oriented graphite impurities or to other paramagnetic impurities embedded in CNTs. The analogies in the angular dependence of the resonance line position and of the resonance line width of these 2 lines suggest that the uncoupled electron responsible for the second line feels the effect of the electric crystalline field of CNT. Hence this impurity should be either located on CNT on in its immediate vicinity.

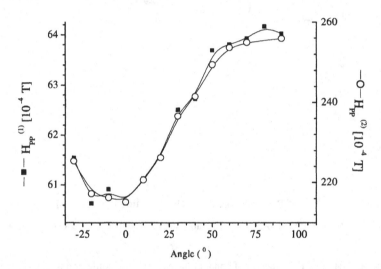

Figure 2. The angular dependence of the resonance linewidth for the two lines that contributes to the central ESR spectrum of CNF.

We suggest that the narrow line originates from electron delocalized over carbon nanotubes and that are not in interaction with the magnetic impurities. Electrons delocalized over regions of nanotubes where magnetic impurities are present (such as the tip of carbon nanotubes) are responsible for the broad resonance.

CONCLUSIONS

ESR investigations on CNF have been reported. The resonance spectrum has been deconvoluted into two main components, a faint line located on the low field wing of the main resonance assigned to magnetic impurities and in a convolution of two asymmetric lines originating from electrons delocalized over conducting domains. One of these lines (the broadest one) has been assigned to electrons delocalized over the conducting domains of CNTs. The origin of the other line is still under investigation. Previous experiments showed that Raman spectroscopy [8, 9] and X-Ray diffraction [9], may be used to assess the orientation of nanotubes. This contribution proves for the first time that ESR spectroscopy may be used to assess the orientation of CNF.

ACKNOWLEDGEMENTS

This research has been funded by US ARMY STTR #A2-1299 grant awarded to the Chemistry Department of Indiana University (Bloomington) and to PartTec Ltd. and by Rice University

REFERENCES

1. Ruan, S.L.; Gao, P.; Yang, X.G.; Yu, T.X., *Polymer*, 2003, *44*, 5643.

2. Park, C.; Ounaies, Z.; Watson, K.A.; Crooks, R.E.; Smith, J.Jr.; Lowther, S.E.; Connell, J.W.; Siochi, E.J.; Harrison, J.S.; StClair, T.L., *Chem. Phys. Lett.*, 2002, *364*, 303.

3. Bai, J.B.; Allaoui, A., *Composites A*, 2003, *34*, 689.

4. Singer, L.S.; Wagoner, G., *J. Chem. Phys.*, 1962, *37*, 1812.

5. Huber, D.L.; Urbano, R.R.; Sercheli, M.S.; Rettori, C *Phys. Rev. B*, 2004, *70*, 125417.

5. Feher, G.; Kip, A.F., *Phys. Rev.*, 1955, *98*, 337.

6. Zhou, W.; Vavro, J.; Guthy, C.; Winey, K. I.; Fischer, J.E.; Ericson, L .M.; Ramesh, S.; Saini, R.; Davis, V.A.; Kitrell, C.; Pasquali, M.; Hauge, R.H; Smalley, R. E., *J. Appl. Phys.*, 2004, *95*, 649.

7.   Kumar, S.; Dang, T.D.; Arnold, F.E., Bhattacharyya, A.R.; Min, B.G.; Zhang, X.; Vaia, R.A.; Park C.; Adams, W.W.; Hauge, R.H., Smalley, R.E.; Ramesh S.; Willis, P.A., Macromolecules, 2002, 35, 9030

Mater. Res. Soc. Symp. Proc. Vol. 887 © 2006 Materials Research Society   0887-Q10-05

Formation, evolution and degradation of nanostructured covalent thin films deposited by Low Energy Cluster Beam Deposition

Giuseppe Compagnini, Luisa D'Urso, A.Alessandro Scalisi, Corinna Altamore, Orazio Puglisi

Dipartimento di Scienze Chimiche, Università di Catania, Viale A. Doria 6 Catania 95125 Italy

ABSTRACT

Low Energy Cluster Beam Deposition (LECBD) is considered an intriguing technique to obtain thin layers with well defined structures at the nano- and meso-scale levels, allowing novel optical, electronic and magnetic properties. The produced layers are highly porous and extremely reactive due to the high surface to volume ratio and must be characterized with "in situ" techniques in order to study their original composition and their evolution once exposed to reactive gases. In this work we present a general overview and some results on the formation evolution and deposition of silicon and carbon cluster beams produced using a laser vaporization source.

INTRODUCTION

The study of small atomic clusters is considered one of the most important and active areas in modern material science for several reasons. Indeed atomic clusters are applied in catalysis as well as in nanoelectronics and many other fields[1-3]. In this frame the production of supersonic cluster beams by using laser ablation sources is frequently used since the works by Smalley et al. where they suggested the possibility to obtain cluster beams even from the most refractory materials[4]. The method has also lead to the discovery of fullerenes and can be applied to a large number of atomic species[5,6], handled in order to obtain a large variety of novel materials. Moreover ultra-cold collision-less supersonic beams are used to study highly reactive species in many aspects. When seeded supersonic beams are deposited to form a so called "cluster assembled layer" the process is generally termed Low-Energy Cluster Beam Deposition (LECBD) because the films are built by the deposition of preformed, size controlled clusters at low kinetic energy (0.1-1 eV/atoms). In this case clusters reach the substrate without consistent fragmentation and retaining their original structure even in the solid state. The obtained materials are highly porous with densities as low as about one half of the corresponding bulk densities. The characteristic nanostructured morphology and the possible memory effect of the original free cluster structures are at the origin of their specific properties. Many examples of novel materials obtained by LECBD are offered in literature. For instance the formation of transition metal cluster assembled films with specific magnetic behaviour resulting from the competition between the intrinsic properties of the grains and the interactions between grains [7], or the deposition of metal oxides with interesting optical and thermal properties[8]. Here we restrict our interest to covalent materials such as carbon and silicon, showing that the formed cluster assembled layers posses features far from those obtained using other deposition techniques.

EXPERIMENTAL DETAILS AND CLUSTER BEAM FEATURES

Cluster beams were generated by using a laser vaporization source similar to that reported by A.Perez et al.[9]. Briefly, a plasma created by the impact of a Nd:YAG laser beam (λ=532 nm, pulse duration=5 ns, frequency=10 Hz) onto a rod was generated in a small chamber and thermalised by the injection of a pulsed helium stream at high pressure (2-10 bar) which permits the cluster growth. Carbon and silicon plumes were generated using polycrystalline graphite and silicon rods (99,99% of purity) respectively. A huge pressure gradient produces a supersonic expansion that takes place at the nozzle exit of the source, where the pumping system ensures a background pressure of 10^{-7} mbar. Naturally ionized clusters, are detected by a so called "orthogonal" Time-Of-Flight Mass Spectrometer (o-TOF-MS), built up according to the Wiley-McLauren set up[10]. In such a system, charged particles are extracted at 90° with respect to their initial velocity and then accelerated and injected into a 1.5 m field free flight tube. Immediately after the acceleration and before the free flight region, two deflection plates adjust the beam trajectory in such a way to strike the Micro-Channel Plate (MCP) detector positioned at the end of the tube. Acquisition is obtained by fast transient recording with a 1 GHz analyzer.

A fine beam of clusters is formed and deposited (thicknesses 100-500 nm) with a rate of about 20 Å/min, on several substrates, situated at a distance of 0.8 m from the expansion nozzle. These substrates were suitably cleaned before their use and kept at room temperature during the depositions in a chamber with a base pressure of 10^{-9} mbar. An overall sketch of the system is shown in fig.1.

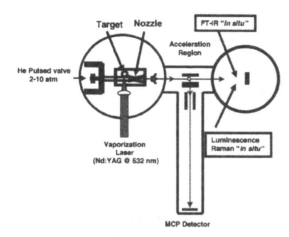

Figure 1: experimental setup for a LECBD deposition

Immediately after the deposition the samples have been positioned either in front of a fused silica window to perform "in situ" Raman scattering experiments or inside a tube, flanged with the UHV chamber, to perform "in situ" infrared absorption spectroscopy through two KBr windows. Raman scattering have been excited either by a 633 nm radiation from a He-Ne laser or by a 514.5 nm one from an Ar ion laser. The backscattered light was collected by suitable optics and analysed by a confocal system. A Jobin Yvon single monochromator equipped with a notch filter and a CCD detector cooled at liquid nitrogen temperature (3 cm^{-1} spectral resolution) is used. With the same Raman apparatus, luminescence can be excited by the 514.5 nm radiation and analysed. FTIR spectra have been collected using a Bruker Equinox 55 spectrometer with a resolution of 1 cm^{-1}. In order to investigate the morphological features of the deposited clusters, the above samples were inspected by Scanning Force Microscopy (SFM). SFM was carried out in air using a commercial instrument (Nanoscope IIIa, Digital Instruments) equipped with commercially available etched silicon probes having a pyramidal shape tip with a nominal curvature and a nominal internal angle of 10 nm and 35°, respectively.

Fig.2 reports some specific features of a typical silicon cluster beam produced. The figure indicates the two most important properties achieved with the machine and concerning the cluster beam, that is the mass distribution and the kinetic energy distribution considered cluster by cluster. In the case shown, the mass distribution is centered between 10 and 40 silicon atoms per cluster and it is independent by the type of ions detected (anions or cations, not shown in the figure). Due to the

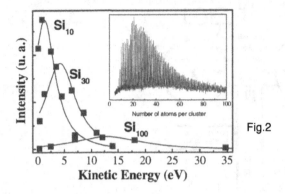

Figure 2: Mass and kinetic energy distributions for silicon clusters seeded in the He beam

moderate growth conditions achieved in our source, no "magic numbers" have been found in any of the recorded spectra. Note that in the mass spectrum the ions belonging to the series $Si_nC^\pm$ or $Si_nO^\pm$ are absent so that we can neglect the presence of oxygen and carbon contaminations. Regarding hydrogenation, actual instrumental resolution is not able to separate hydrogenated clusters as well as isotopic shifts, at least for medium and large size clusters.

Since one of the major features of LECBD technique is the possibility to deposit thin films starting from aggregates which propagate with a low kinetic energy, we measured carefully the kinetic energy distribution of the aggregates[11]. It has been found that silicon aggregates obtained with our source posses speed distributions centered at around 900 m/sec with a width (FWHM) of 400 m/sec. Note that in free jets' cluster beam sources all the produced aggregates may tend to have the same ion speed rather than the same ion energy, so that this speed value can be considered as the speed of all the produced species (ionized and neutral). Evaluated energies are around 0.1-0.2 eV/atom, far below the impact fragmentation limit[12].

DEPOSITION AND DEGRADATION OF SILICON CLUSTER ASSEMBLED FILMS

The conversion of the mass distribution into a size distribution is hard to be estimated because of the unknown geometry. However, in the case of silicon, it has been observed that whether in the case of cage-like structures or in the case of compact sp^3 hybridized structures, the average bond length is not far from the bulk value (2.3 Å) and then the use of the bulk silicon atomic density seems to be a reasonable guess[13]. Fig.3 reports the cluster size distribution as obtained from the already reported mass spectrum, in the hypothesis of a spherical shape. It is straightforward to individuate average cluster sizes in the range 0.8-1.5 nm. The same fig.3 compares such a distribution with that obtained after cluster landing by performing SFM analysis in the case of a sub-monolayer coverage. A HOPG substrate has been chosen because of the known low coalescence of silicon aggregates onto such surfaces. The two distributions are in general agreement even though a high cluster size tail of the distribution is observed in the solid phase where higher diameters are observed of more than 2 nm. We can ascribe the result either to a coalescence after landing or to a scarce sensitivity of the TOF system to detect bigger aggregates. The partial overlap of the gas phase and solid phase size distributions suggests that particles land on the substrate at least in part maintaining their features.

Careful studies on the bonding state of deposited silicon cluster beams have been first conducted by J.E.Bower et al.[6] using photoelectron spectroscopy. They showed that oxygen uptake on the deposited clusters has an initial sticking coefficient substantially smaller that on bulk silicon but comparable to the one found for larger clusters. It was also observed that O/Si saturation ratios remains below 1 after exposing the clusters to 10^5 Langmuir. These data are partially in agreement with our recent experiments performed by "in situ" infrared spectroscopy. There we observed that

Fig.3

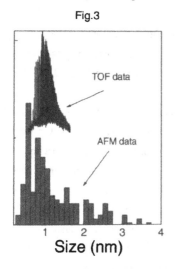

Size (nm)

Figure 3: Comparison between mass spectroscopy and AFM data regarding the cluster size distribution for silicon aggregates before and after landing.

oxygen stoichiometry can approach to 2 even though for higher oxygen exposition and with a very slow kinetics. Specifically, cluster assembled silicon layers have been deposited onto suitable KBr substrate, allowing transmission experiments in the 600-3000 cm^{-1} spectral range. Two regions are present with interesting features, related to the chemical state of the silicon clusters. The first is located between 600 and 1400 cm^{-1}. In this range a number of signals coming from various Si-O stretching and bending modes together with some Si-H vibrations are found. The second is located between 1900 and 2400 cm^{-1}, where only Si-H stretching modes are detected. General considerations[14] indicate that a broad signal between 800 and 1100 cm^{-1} can be attributed to the presence of Si-O-Si vibrations in a SiO$_x$ samples. The band appears at 1080 cm^{-1} in amorphous SiO$_2$ and is red shifted when the oxide is sub-stoichiometric following a trend depending on the nature of the deposited layer. Moreover an Si-O-Si bending mode is found at around 800 cm^{-1}.

Fig.4 shows infrared absorption spectra taken at different oxidation stages in the range 600-1400 cm^{-1}. Immediately after the deposition we observe a band at 780 cm^{-1} and a broad weak contribution located between 800 and 1050 cm^{-1}. The first one can be attributed to SiH$_x$, where silicon atoms are partially back-bonded to oxygen atoms because of a weak oxidation of the hydrogenated silicon system, while the broad signal (800-1050 cm^{-1}) confirms the presence of a small quantity of Si-O bonds. Photoluminescence has been measured for such a sample and a low spectral intensity is obtained (see the same fig.4).

Once the sample is exposed to atmospheric pressure the situation gradually changes. After few minutes an intense signal (extending from 750 cm^{-1} to 1100 cm^{-1}) is evidenced. Beside the above mentioned H-Si(Si$_2$O) mode, this structure contains a large variety of oxide species (SiO$_x$) and the red-shift indicates an increasing oxidation.

There is also an evolution of the Si-H stretchings with more complex spectral features observed and a general blue-shift of the original signal (not shown here). This last effect can be due to two main reasons. The first is the bonding of electronegative elements (oxygen) which withdraw electron density from the silicon atoms and subsequently decrease the Si-H bond length increasing the Si-H vibrational frequency[15]. The second is related to the evolution of SiH$_2$ species to SiH ones. In any case the overall Si-H stretching intensity progressively decreases with increasing oxidation.

It is interesting to observe that a PL signal located between 600 and 800 nm is progressively enhanced by oxidation reaching a maximum intensity after several minutes. This effect, due to recombination centers at the Si/oxide interface, is in agreement with the quantum Confinement/Luminescent Center model, recently supported by a number of experiments[16]. In this case electron-hole pairs generated in the silicon clusters (having a

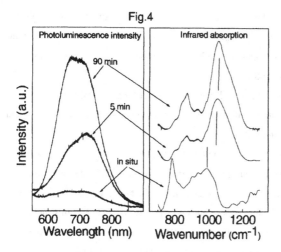

Fig.4

Photoluminescence intensity | Infrared absorption

90 min

5 min

in situ

Intensity (a.u.)

600 700 800
Wavelength (nm)

800 1000 1200
Wavenumber (cm⁻¹)

Figure 4: IR and photoluminescence spectra for silicon cluster assembled films as a function of their ageing time in atmosphere.

quantum-size enhanced bandgap) radiatively recombine through luminescent centers located at the Si/oxide interface or within the surrounding oxide layer. Thus the PL emission of a given sample should strongly depend on the nature of surface passivation. Moreover, following the most recent highlights[17], the observed luminescence should come from particle sizes in the range 1.5-3 nm. In this frame it seems that smaller particles found in our cluster assembled films do not contribute to the luminescence spectra.

DEPOSITION AND DEGRADATION OF CARBON CLUSTER ASEMBLED FILMS

Nanostructured carbon films can be obtained by using several deposition methods and a large number of studies and results are present in literature[18,19]. In most of these works the carbon network is generally composed by a large number of sp^2 bonded carbons in which a gallery of carbon nanospecies are found: from nanotubes and fullerenes to carbon onions and diamond nanoaggregates. These species are generally tailored by the deposition methods, by the substrate used and by the presence of foreign species in the deposition system (see for instance the catalytic effect of Fe and Ni in the production of carbon nanotubes). An intriguing alternative to generate cluster assembled carbon thin films is the use of a laser vaporization source which generates a carbon plasma within the throat of a high pressure pulsed nozzle. This method has been widely investigated by A.Rohlfing et al.[5] and has been found to form linear carbon clusters as well as carbon rings in a large extent. Following the deposition, these species partially retain their bonding state, as it was suggested by vibrational analysis, leading to a carbon mixture in which both sp^2 and sp hybridized carbon atoms are present[20].

While the sp^2 component is essentially stable under ambient temperature and pressure, the carbynoid part is extremely unstable in both its two components (polycumulene and polyyne) and evolves towards the more stable graphite-like state.

Once again vibrational spectroscopies constitute the best techniques to detect these species and to follow their degradation. In particular Raman spectra are generally full of information because the position and relative intensities of the bands are indicative about the bonding state of the carbon atoms and also because the shape and width of the bands are directly connected with structural relaxation phenomena or with the extension of the nanocrystalline domains which are typically found in the amorphous microscopic structure. Typical infrared and Raman signals coming from a cluster assembled carbon film obtained with the LECBD technique is shown in the inset fig.5. Two prominent broad signals located in the ranges 1000-1600 and 1800-2300 cm^{-1} are clearly visible and can be attributed to the sp^2 and sp components respectively. The graphite-like feature is generally considered as the convolution of a D band (1350 cm^{-1}) and a G one (1550 cm^{-1}), while the sp signal should contain both those of polycumulenes (lower Raman shifts) and polyynes (higher Raman shifts) species. The same fig.5 shows the evolution of the G band position and the decrease of the sp band intensity (with respect to the sp^2 signal considered as I_{DG}) when the deposited sample is exposed to atmosphere. Recent studies observed this degradation also in the case of the interaction of the produced films with unreactive gases like He, Ar or N_2[20]. It seems that while oxygen chemically reacts with polyynes and polycumulenes causing an almost complete degradation of these species, the effect of unreactive gases is merely mechanic with the gas species hitting the sp chains and causing the degradation. Regarding the nature of this degradation and following a general literature consensus[21], the blue shift of the G band (correlated to the decrease of the carbynoid component) can be attributed to the transformation of linear sp chains into ordered hexagonal sp^2 graphitic domains. This is also confirmed by an increase of the I_D/I_G intensity ratio. Particularly, the increase of this quantity can be directly connected with the extension (L) of the graphitic domains (better to say with their correlation length) through the relation[21]: $I_D/I_G=AL^2$. Here A is a constant which depends on the excitation wavelength. In the hypothesis of a transformation of the carbynoid component into hexagonal graphitic domains, the increase of I_D/I_G

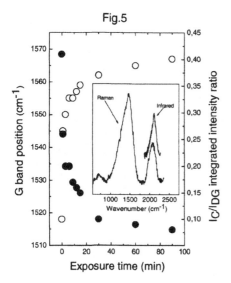

Fig.5

Figure 5: G band position and normalized carbyne signal intensity as a function of the exposure time in air

and the decrease of I_C/I_{DG} should be strongly correlated if one consider that: $I_C/I_{DG}=B(N_{sp}/N_{sp2})$, where B is a scattering cross section ratio while N_{sp} and N_{sp2} are the concentrations of carbon triple and double bonds respectively.

It is the straightforward to obtain an inverse linear correlation between I_D/I_G and I_C/I_{DG} and this has been found as shown in fig.6. In the same figure the result of a linear fitting procedure has been shown following the relation:

$$\left(\frac{I_D}{I_G}\right)^{-1} = \alpha\left(\frac{I_C}{I_{DG}}\right) + \beta \tag{1}$$

The fitting parameters α and β are directly related with the Raman scattering cross section ratio B and with the constant (A) previously considered. Then in our case we find that

$$\frac{I_C}{I_{DG}} = 0.6\frac{N_{sp}}{N_{sp2}} \tag{2}$$

which allow to evaluate directly the relative concentration of sp and sp^2 bonds as reported in the same fig.6 (top scale).

Other ways of degradation of LECBD films have been studied. Moderate thermal heating up to (200 °C) strongly affect the degradation of the sp carbon chains leaving the graphite-like part almost

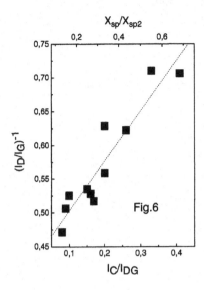

Figure 6: Correlation between I_D/I_G and I_C/I_{DG} as reported by eq.1. Top scale gives quantitatively the sp/sp^2 ratio.

unchanged, while higher temperature thermal annealings also modify the latter towards an increase in graphitization.

Regarding the irradiation with energetic particles it has been found that degradation strongly involve all the structures in a rather unusual way. In fact, irradiation by means of 200 keV Ar$^+$ beams produces effects on both the graphite-like and the sp signals. The first is modified by changing the D/G intensity ratio and the G signal position as already reported[22], while the sp signal decreases its intensity by increasing the ion fluence. Even though a mere observation of the Raman signal support similar considerations as those already given by degrading sp carbon chains by a simple thermal treatment or by gas exposition, an "in situ" study performed by infrared transmission spectroscopy has shown that a new sharp component at 2110 cm^{-1} appears, correlated to the above mentioned broad band decreases with increasing fluence[23]. This has been shown in fig.7 as an inset. The rest of fig.7 gives quantitatively the decrease of the wide component and the increase of the sharp one as a function of the ion fluence (φ). These data are obtained by a proper deconvolution of the overall signal into a sharp component (2110 cm⁻¹) and two contributions,

Fig.7

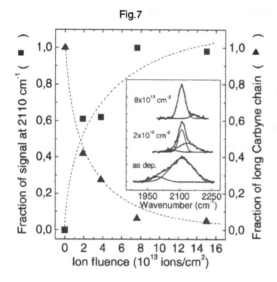

Figure 7: Evolution of the 2110 cm⁻¹ line and the broad triple bond signal as a function of the ion fluence. The inset shows IR spectra (absorbance) in the 1900-2250 cm⁻¹ region where these signals are located.

related to the broad band (polycumulene and polyyne). It is interesting to observe that fits of the reported experimental data with exponential curves such as:

(i) $1-e^{-\sigma\varphi}$ (growth of the 2110 cm⁻¹ sharp line)

(ii) $e^{-\sigma\varphi}$ (decrease of the broad triple bond stretching band)

give similar cross sections (σ). In particular these are $3.7\pm1 \times 10^{-14}$ cm² for the former and $4.3\pm1\times10^{-14}$ cm² for the latter, suggesting a correlation between the two phenomena. Similar analysis on the already reported Raman data gives cross sections five times greater. This discrepancy can be attributed to differences in the processes involved to evaluate the vibrational spectrum (that is differences in the spectroscopic process between Raman scattering and infrared absorption).

Following a wide consensus present in literature, it is easy to attribute the sharp 2110 cm⁻¹ signal to the formation of monosubstituted acetylenic, diacetylenic or oligoacetylenic structures[24] which are formed upon irradiation. The similarity between the two cross sections obtained, confirms that these structures are formed as a consequence of the fragmentation of the long chains, while the decrease of the signal in the Raman spectrum can be due to the formation of polar molecular structures, more easily revealed with IR spectroscopy than with Raman scattering (mutual exclusion rule).

REFERENCES

1.A.W.Castleman,K.H.Bowen, J.Chem.Phys. **100**, 12911 (1996)

2.W.Eberhard Surf.Sci. **500**, 242 (2002)

3.C.Binns Surf.Sci.Rep. **44**, 1 (2001)

4.see for instance J.B.Hopkins, P.R.R. Langridge, M.D.Morse, R.E.Smalley, J.Chem.Phys. **78**, 1627 (1983)

5.R.A.Rohlfing, D.M.Cox, A.Kaldor, J.Chem.Phys. **81**, 3322 (1984)

6.J.E.Bower, M.F.Jarrod, J.Chem.Phys. **97**, 8212 (1992)

7.V.Dupuis, L.Favre, S.Stanescu, J.Tuaillon-Combes, E.Bernstein, A.Perez J.Phys.Cond. Matter **16**, S2231 (2004)

8.P.Siciliano, A.M.Taurino, T.Toccoli, A.Pallaoro, S.Iannotta, P.Milani, Chem.Sensors **20**, 368 (2004)

9.A.Perez, P.Melinon, V.Dupuis, P.Jensen, B.Prevel, J.Tuaillon, et al. J.Phys. D **30**, 709 (1997)

10. W.C.Wiley and I.H.McLaren, Rev.Sci.Instr. **26**, 1150 (1955)

11. G.Compagnini, L.D'Urso, O.Puglisi Mat.Sci.Eng. (in press, 2005)

12. R.Neuendorf, R.E.Palmer, R.Smith, Chem.Phys.Lett., **333** (2001) 304

13. R.O.Jones, B.W.Clare and P.J.Jennings, Phys.Rev. B, **64** (2002) 125203

14. M.H.Brodsky, M.Cardona, J.J.Cuomo, Phys.Rev. B, **16** (1977) 3556

15. P.Gupta, A.c:Dillon, A.S.Braker, S.M.George, Surf.Sci., **245** (1991) 360

16. S.M.Prokes, O.J.Glembocki, Phys.Rev. B, **49** (1994) 2238

17. G.Ledoux, O.Guillois, D.Porterat, C.Reynaud, F.Huisken, et al. Phys.Rev., **B62** (2000) 15942

18. S.Iijima, Nature **354**, 56-58(1991)

19. E. Koudoumas, O. Kokkinaki , M. Konstantaki, S. Couris, S. Korovin, P. Detkov, V. Kuznetsov, S. Pimenov, V. Pustovoi, Chem.Phys.Lett. **357**, 336 (2002)

20. C.S.Casari, A.LiBassi, L.Ravagnan, F.Siviero, C.Lenardi, et al. Phys.Rev. B **69**, 075244 (2004)

21. A.C.Ferrari and J.Robertson Phys.Rev. B **61**, 14095 (2000)

22. G.Compagnini Mat.Res.Soc.Symp.Proc.**851**, 41 (2005)

23. G.Compagnini, S.Battiato, O.Puglisi, G.A.Baratta, G.Strazzulla, Carbon **43**, 3025 (2005)

24. Bellamy LJ. The Infra-red Spectra of Complex Molecules. Chapman and Hall London 1975

Degradation Processes in
Polymer-Nanofiller Composites

Mater. Res. Soc. Symp. Proc. Vol. 887 © 2006 Materials Research Society   0887-Q01-02

Multi-scale Rule-of-Mixtures Model of Carbon Nanotube/Carbon Fiber/Epoxy Lamina

S. J. V. Frankland[1], J. C. Riddick[2] and T. S. Gates[3]
[1]National Institute of Aerospace, 100 Exploration Way, Hampton, VA 23666, U. S. A.
[2]Vehicle Technology Directorate, US Army Research Laboratory, Hampton, VA 23681, U.S.A.
[3]Mechanics of Structures and Materials Branch, NASA Langley Research Center, Hampton, VA 23681, U. S. A.

ABSTRACT

A unidirectional carbon fiber/epoxy lamina in which the carbon fibers are coated with single-walled carbon nanotubes is modeled with a multi-scale method, the atomistically informed rule-of-mixtures. This multi-scale model is designed to include the effect of the carbon nanotubes on the constitutive properties of the lamina. It included concepts from the molecular dynamics/equivalent continuum methods, micromechanics, and the strength of materials. Within the model both the nanotube volume fraction and nanotube distribution were varied. It was found that for a lamina with 60% carbon fiber volume fraction, the Young's modulus in the fiber direction varied with changes in the nanotube distribution, from 138.8 to 140 GPa with nanotube volume fractions ranging from 0.0001 to 0.0125. The presence of nanotube near the surface of the carbon fiber is therefore expected to have a small, but positive, effect on the constitutive properties of the lamina.

INTRODUCTION

Near term applications of carbon nanotubes (CNT) in materials for aerospace vehicles are most likely to be realized with mechanical or electrical applications that require small quantities of nanotubes. To implement carbon nanotubes as structural members could be accomplished by the selective use of carbon nanotubes in critical areas of traditional composites. For example, carbon nanotubes added between the layers of a traditional carbon fiber/epoxy laminate may well improve the interface strength between carbon fiber and epoxy matrix, and thereby improve fracture toughness of these laminates.

The effectiveness of adding carbon nanotubes to laminates can be analyzed with multi-scale models which are capable of incorporating atomistic details. At the atomistic-level the assumptions of 'perfect' bonding between the various contributors to the laminate can be removed in detailed atomistic simulations. With properties at this level established, the micro-scale properties can be accessed with more standard micromechanics methods such as the Mori-Tanaka method or the rule-of-mixtures analysis[1]. Multi-scale models have applied atomistic simulation and micromechanics to the constitutive properties of various functionalized nanotube materials [2-5]. There are some atomistic simulations of epoxy/nanotube composites by other researchers which address nanotube pull-out [6]. Also, a more recent study on nanotubes chemically bonded into the epoxy matrix showed a

Young's modulus of up to 160 GPa in the direction of the nanotube axis, and 4-8 GPa in the transverse direction at a nanotube volume fraction of 25 % [7].

In the present work, the constitutive properties of a unidirectional carbon fiber/epoxy lamina are modeled by a multi-scale model. In the lamina, each carbon fiber is uniformly coated with carbon nanotubes. The objective is to calculate the Young's modulus in the direction of the carbon fiber of unidirectionally-reinforced carbon fiber/epoxy laminae, where the nanotube distribution around the fiber is varied. For this purpose, we develop the atomistically informed rule-of-mixtures approach. This method requires a multi-scale approach to the analysis of several components of the nanotube-coated carbon fiber. The constitutive properties of the nanoscale components are modeled with the molecular dynamics/equivalent continuum (MD/EC) model. The CNT distribution and volume are modeled using the MD/EC model in conjunction with the Mori-Tanaka micromechanics method to calculate the Young's modulus over a defined nanotube distribution. A description of each stage of the multi-scale model is presented, and the results of applying the atomistically-informed rule-of-mixtures are given for a lamina with 60% carbon fiber volume fraction.

MATERIAL DESCRIPTION

A unidirectional carbon fiber-reinforced polymer lamina [8] (Fig. 1) is constructed such that the carbon fibers are coated with carbon nanotubes (CNT) and then cured in epoxy. The schematic in Fig. 1 defines the principal material axes with respect to the fiber direction. Herein, the 1-, 2-, and 3-axes, are referred to as longitudinal, transverse, and out-of-plane, respectively. The epoxy was comprised of bisphenol F resin reacted with epichlorohydrin and cured with triethylenetetramine (TETA). The carbon nanotubes were (10,10) single-walled carbon nanotubes of radius 0.678 nm. It is assumed in the model that there is no chemical bonding between the 3 components: carbon fiber, carbon nanotubes, and the epoxy.

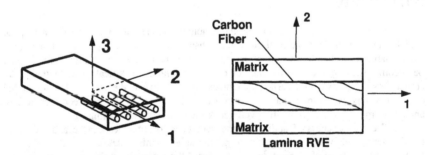

Figure 1. Unidirectionally-reinforced lamina

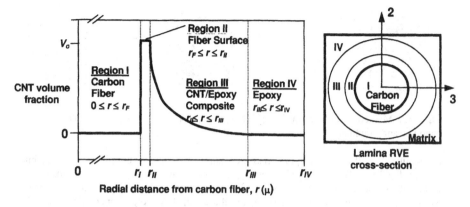

Figure 2. Details of variations of CNT volume fraction in the representative volume element for a lamina with CNT-coated unidirectional carbon fibers

A typical representative volume element (RVE) is shown for a unidirectionally-reinforced lamina in Fig. 1. Details of the variations in the micro-structure resulting from coating the unidirectional carbon fibers with CNT are depicted in Fig. 2, along with details of the lamina RVE cross-section (1-axis, positive out-of-the-page). The carbon fiber and its surroundings are represented by four regions. The CNT volume fraction versus radial distance from the carbon fiber for each region is shown in Fig. 2. Region I is the carbon fiber itself, where r_F represents the carbon fiber radius. Region II represents the carbon fiber surface, and it has three components: the graphitic surface of the carbon fiber, carbon nanotubes and epoxy, and ranges from the carbon fiber radius at r_F to r_{II}, 0.004 microns from the carbon fiber. The CNT volume fraction in Region II is defined by the parameter V_o, shown in Fig. 2. Region III is the region surrounding the carbon fiber which is comprised of carbon nanotubes and epoxy. As depicted in Fig. 2, the CNT volume fraction in this region decays, from V_o to zero, as the radial distance from the carbon fiber increases. The width of Region III ranges from r_{II} to r_{III}, and is denoted by the parameter Δ, where $\Delta = r_{III} - r_{II}$. Region IV is the remaining epoxy matrix, and ranges from r_{III} to a maximum width denoted by parameter r_{IV}.

MULTI-SCALE MODEL

Molecular dynamics simulations

Representative volume elements (RVEs) of the molecular structure of the material system were generated via molecular dynamics (MD) simulation. In particular, three different systems with simulated with MD: the epoxy matrix (Fig. 3(a)), the epoxy matrix containing

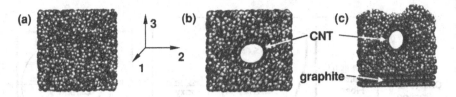

Figure 3. Molecular RVEs of (a) Epoxy, (b) epoxy nanotube composite, and (c) epoxy nanotube and graphite.

a nanotube (Fig. 3b), and the epoxy matrix containing a nanotube, but placed on a graphite surface (Fig 3(c)). The epoxy and the epoxy/nanotube were periodic in all three dimensions, and the three-component system was periodic in the 1- and 2- directions. In the 3-direction was periodic, but extended to 1 micron, and is, therefore, effectively an open boundary.

All three systems were simulated with the AMBER force field [9] using the LAMMPS [10] and DL-POLY [11] simulation codes. All the nanotube and graphite parameters were taken from the AMBER force field. Several of the force field parameters of the epoxy network were derived from *ab initio* calculations carried out with the NWChem package [12]. The partial atomic charges were derived from an RESP fit to a Hartree-Fock calculation of wavefunction. The bond lengths, angles, and dihedrals, were calculated by optimizing the geometry of parts of the epoxy network at the HF/STO-3G level. Partial charges were then fit at this geometry from a HF/6-31G* calculation. The rest of the parameters, were taken from the AMBER force field. In the simulations, electrostatic interactions were calculated with Coulomb's law, and an Ewald summation was used for the long range Coulombic interactions. Electrostatic interactions and 1-4 non-bonded dihedral interactions were scaled by the standard AMBER parameters of 0.83 and 0.50 respectively.

The molecular structure of the epoxy was a network of bisphenol F resin molecules reacted with epichlorohydrin and cross-linked via triethylenetetramine (TETA) molecules. The TETA molecule is able to cross-link with up to four resin chains, and on average 2.3 of the 4 TETA amine groups in the resultant structure were involved in cross-linking. The equilibrium structures were obtained by compressing the epoxy network until the minimum potential energy configuration at finite temperature (300 K) was reached. In the 3-component system, the epoxy was prepared with periodicity in the 1- and 2- directions, and was compressed between graphite plates. The upper graphite plate was then removed from the simulation, and the system was further equilibrated to minimize the potential energy.

Equivalent-Continuum Model

The elastic constants were then obtained for each of the molecular RVEs from equating the energies of deformation for the molecular system, which were calculated from molecular dynamics simulation, to the energies of deformation of an equivalent continuum solid under identical deformations. The material constitutive relation of stress σ and strain ε is given by Hooke's law

$$\sigma = \mathbf{C}\varepsilon. \tag{1}$$

The stiffness matrix C of an orthotropic system has 9 components:

$$\mathbf{C} = \begin{pmatrix} C_{11} & C_{12} & C_{13} & 0 & 0 & 0 \\ C_{12} & C_{22} & C_{23} & 0 & 0 & 0 \\ C_{13} & C_{23} & C_{33} & 0 & 0 & 0 \\ 0 & 0 & 0 & G_{23} & 0 & 0 \\ 0 & 0 & 0 & 0 & G_{13} & 0 \\ 0 & 0 & 0 & 0 & 0 & G_{12} \end{pmatrix} \tag{2}$$

The 9 components can be determined from the boundary conditions given in Ref. [2]. As in previous work [2], the axial stiffness C_{ii} and shear moduli G_{ij} were determined directly, and the stiffnesses C_{ij} were calculated from the plane strain bulk moduli K_{ij} and C_{ii}. For these cases, i and j are any of the 1-, 2- or 3- principal directions. The C_{ii} and K_{ij} were determined for all three systems. For the epoxy/nanotube system all nine constants were computed. The shear moduli of the remaining two systems were not needed for the model.

In the MD simulations, the displacement fields from Ref. [2] were applied to the each RVE structure in strain increments of 0.0025 every 10,000 steps at 1 fs each (10 ps per increment) up to a maximum strain of greater than 0.02. For the epoxy and epoxy/nanotube systems, application of the displacement field included displacing both the periodic boundaries and the atomic positions, then equilibrating the system with MD to its new state point. For the three-component system, the upper layer of the epoxy and the lower graphite sheet were fixed in location. The atoms in the fixed layers were displaced, but no further equilibration was performed on them.

Because of the limitation of rectilinear boxes in LAMMPS, DL-POLY was used for the shear moduli. Care was taken to translate the force field as accurately as possible between the codes, but some variation in the results is still expected.

Mori-Tanaka Micromechanics

The elastic constants determined for the molecular RVEs with the molecular dynamics/equivalent continuum method described above, are then used as input to the Mori-Tanaka micromechanics method to calculate the properties of an embedded epoxy/nanotube effective fiber in an epoxy matrix [13-15]. The stiffness tensor **C** of this nanocomposite as a function of effective fiber volume fraction v_f is

$$\mathbf{C} = \left((1 - v_f)\mathbf{C}^m + v_f \mathbf{C}^f \mathbf{T}^f \right) \left((1 - v_f)\mathbf{I} + v_f \mathbf{T}^f \right)^{-1} \tag{3}$$

where the dilute strain concentration tensor $\mathbf{T}^f$ is

$$\mathbf{T}^f = \left[\mathbf{I} + \mathbf{S}^f \left(\mathbf{C}^m \right)^{-1} \left(\mathbf{C}^f - \mathbf{C}^m \right) \right]^{-1} \tag{4}$$

and S^f is the Eshelby tensor for a prolate ellipsoid [16] with aspect ratio 1:1000. The superscripts and subscripts f and m refer to the effective fiber and matrix properties, respectively, and I is the identity tensor. The v_f is related to the nanotube volume fraction v_{NT} as

$$v_{NT} = v_{NT,RVE} v_f \tag{5}$$

where $v_{NT,RVE}$ is the volume fraction of nanotubes in the molecular RVE. In equation (3) the stiffness matrix C is calculated for aligned effective fibers. To randomize the orientation of the effective fibers in the nanocomposite[2,17], and therefore the nanotubes, the components of C are

$$\left\langle C_{ijmn} \right\rangle = \left(\kappa - \frac{2}{3}\mu \right)\left(\delta_{ij}\delta_{mn} \right) + \mu\left(\delta_{is}\delta_{jn} + \delta_{in}\delta_{jm} \right) \tag{6}$$

where $i,j,m,n = 1,2,3$, δ_{ij} is the Kronecker delta and

$$\kappa = \frac{1}{9}C_{iij}$$

$$\mu = \frac{1}{10}\left(C_{ijij} - \frac{1}{3}C_{iijj} \right) \tag{7}$$

With the randomization of the effective fibers, C becomes isotropic.

Distribution of Nanotubes

The Mori-Tanaka analysis yields the modulus of the nanocomposite at different nanotube loadings, but a description of the spatial distribution of the nanotubes around the carbon fiber must also be defined. The nanotube distribution can be described by using the following exponential form:

$$v_d(r,\alpha) = V_o e^{-\alpha(r-r_{II})} - \left(\frac{r-r_{II}}{\Delta} \right) V_o e^{-\alpha\Delta}. \tag{8}$$

where, v_d is the volume fraction distribution; the parameter r measures radial distance from the center of the carbon fiber; V_o is the initial CNT-epoxy volume fraction of the molecular RVE; and Δ denotes the width of the region in which the nanotubes are distributed. In the present analysis, a value for Δ was specified, and the following constraint equation was solved for successive values of the decay parameter α using Mathematica™[18]

$$\sum_{i}^{N} v_d(r_i, \alpha) \int_{r_i}^{r_{i+1}} 2\pi r \, dr = V_N \cdot A. \qquad (9)$$

Here, V_N denotes the lamina volume fraction, A, the lamina RVE surface area in the 2-3 plane, shown in Fig. 2, and N, the number of subregions of Region III. The value of α that satisfies the equality of Eqn (9), for a given value of V_N, is then applied in the relation of Eqn (8) to define the distribution of CNT volume fractions Region III for width Δ.

Atomistically-Informed Rule-of-Mixtures Analysis

The contribution to the longitudinal Young's modulus of the lamina, E, from each of the four regions shown in Fig. (2) is then determined by using an atomistically-informed rule-of-mixtures analysis:

$$E = v_I E_I + v_{II} E_{II} + v_{III} E_{III} + v_{IV} E_{IV}. \qquad (10)$$

where, v_i and E_i are the volume fraction and Young's modulus of each of the 4 regions. For the present analysis, v_I represents the carbon fiber volume fraction, and is fixed at 60 %. The Young's modulus of the carbon fiber, E_I, is 230 GPa. For the present analysis, the volume fraction v_{II} is fixed by setting the value of r_{II} at 0.004 microns. The Young's modulus E_{II} is calculated from the C tensor of the three component molecular RVE, shown in Fig 3c, resulting from the molecular dynamics/equivalent continuum method. The contributions of v_3 and E_3 are variable. In Region III the nanotube distribution varies according to Eqns (8) and (9). The variation of the modulus in Region III is taken from the properties derived by mapping the results of the Mori-Tanaka micromechanics model the CNT volume fraction distributions from Eqns (8) and (9). Finally, the Young's modulus and volume fraction for Region IV, E_{IV} and v_{IV} respectively, are calculated from the C tensor for the epoxy molecular RVE (Fig 3(a)) .

RESULTS AND DISCUSSION

The elastic constants of the molecular RVEs determined from the molecular dynamics simulations and the equivalent continuum model are listed in Table I. The epoxy elastic constants yield a Young's modulus of 1.44 GPa and a Poisson ratio of 0.42. These values are used for the epoxy matrix in the present analysis. Typical values for storage moduli of neat cured epoxy are 2-3 GPa at 250-300 K [19]. The volume fraction of nanotubes in the epoxy/nanotube system was 15.35%, and in the epoxy/nanotube/graphite system, 15.55 %.

Using the nanotube/epoxy composite properties presented in Table I and the properties of the epoxy matrix, the Young's modulus of the nanocomposite with randomized effective epoxy/nanotube fibers is plotted in Fig 4 as a function of nanotube volume fraction. These results are generated using the Mori-Tanaka micromechanics from eqns (3)-(7).

Table I- The Elastic Constants of the Molecular RVEs.

Elastic Constants	Epoxy (GPa)	Epoxy/Nanotube (GPa)	Epoxy/Nanotube/Graphite (GPa)
C_{11}	4.12	99.0	267
C_{22}	5.05	9.43	182
C_{33}	3.60	6.22	9.94
K_{12}	4.78	31.9	138
K_{13}	4.73	32.4	72.2
K_{23}	4.41	12.1	49.5
G_{12}		0.33	
G_{13}		0.32	
G_{23}		0.20	

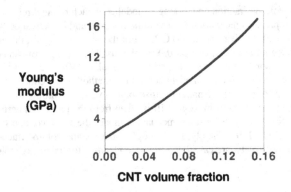

Figure 4. Young's modulus of the epoxy/CNT composite from Mori-Tanaka micromechanics

Fig 5(a) shows variations in CNT volume fraction distributions of exponential form for a lamina RVE with 60% carbon fiber volume fraction and $\Delta = 0.3$ microns. Here, volume fraction variations from Eqns (8) and (9) are depicted for lamina CNT volume fractions ranging from 0.0005 to 0.0055. The exponential form allows the CNT volume fraction distribution to vary from a nearly linear form as shown for $V_N = 0.0055$ through increasingly exponentially varying forms for decreasing values of lamina CNT volume fraction, to a nearly stepwise variation at $V_N = 0.0005$.

Fig 5(b) shows the variations in Young's modulus for Region III. Results are shown for selected nanotube distributions from Fig 5(a). Here, Mori-Tanaka micromechanics have been applied using volume fraction distributions for lamina volume fractions 0.0055, 0.0025, and 0.0005, where $\Delta = 0.3$ microns. The Young's modulus distributions of Fig 5(b) are used directly in the atomistically-informed rule-of-mixtures analysis to represent the contribution of Region III to the constitutive properties of the lamina.

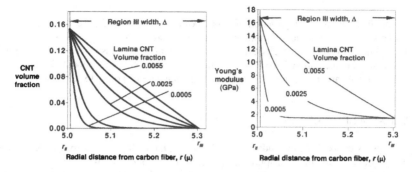

Figure 5. (a) CNT volume fraction distribution in Region III and (b) Young's modulus distribution in Region III, for $\Delta = 0.3$.

Finally, the total longitudinal Young's modulus of the unidirectionally-reinforced lamina with CNT is plotted in Fig. 6 for different sizes of Region III (Fig. 2). For nanotubes within approximately 700 nm of the carbon fiber surface the Young's modulus varies about 0.01 GPa with changes to the size of Region III. The predominant change in the Young's modulus of the coated nanotube fiber is a function of the nanotube volume fraction in the lamina. Up to nanotube loadings of 0.0125, the amount of change in the modulus is 1.2 GPa.

CONCLUSIONS

A multi-scale method, denoted the atomistically-informed rule-of-mixtures, is developed to calculate the Young's modulus of a carbon nanotube coated carbon fiber.

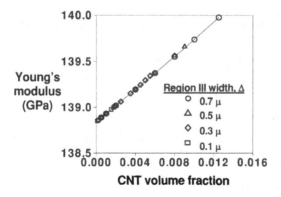

Figure 6. Atomistically informed rule-of-mixtures results for unidirectionally-reinforced lamina with CNT-coated carbon fibers.

Within the method, the distribution of nanotubes in the vicinity of the carbon fiber surface can be varied as a function of nanotube volume fraction. For a lamina with 60% carbon fiber volume fraction, the results indicate that the Young's modulus is less dependent on the local nanotube disribution than on the nanotube volume fraction for nanotube located within 700 nm of the carbon fiber and nanotube loadings in the lamina of up to 1.25 vol %.

ACKNOWLEDGMENTS

The authors thank D. C. Lagoudas, G. D. Seidel, J. Zhu, E. Barrera, P. Thakre, A. Awasthi, and J. Ratcliffe for many helpful discussions. The simulations were carried out at the NASA Ames Supercomputing Center, and on System X at Virginia Polytechnic Institute and State University.

REFERENCES

1. X. L. Gao and S. Mall, J Comp Tech Res 22, (2000)
2. S. J. V. Frankland, G. M. Odegard, and T. S. Gates, AIAA Journal 43, 1828-1835 (2005).
3. T.S. Gates, G.M. Odegard, S.J.V. Frankland and T.C. Clancy, Composite Science and Technology 65, 2416 (2005).
4. G. M. Odegard, S. J. V. Frankland, M. N. Herzog, T. S. Gates, and C. C. Fay. "Constitutive Modeling of Cross-Linked Nanotube Materials" 45th AIAA/ASME/ASCE/AHS/ASC Structures,Structural Dynamics, and Materials Conference, AIAA-2004-1606.
5. T. S. Gates, G. M. Odegard, M. P. Nemeth, and S. J. V. Frankland. "Predicting the Influence of Nano-scale Material Structure on the In-plane Buckling of Orthotropic Plates" 45th AIAA/ASME/ASCE/AHS/ASC Structures, Structural Dynamics, and Materials Conference, AIAA-2004-1607.
6. J. Gou, Z. Liang, C. Zhang, and B. Wang, Composites B 36, 524-533 (2005).
7. T. C. Clancy, and T. S. Gates "Mechanical Properties of Nanostructured Materials Determined through Molecular Modeling Techniques", 46th AIAA/ASME/ASCE/AHS/ACS SDM Conference, AIAA-2005-1852.
8. R. M. Jones, Mechanics of Composites Materials, (Taylor and Francis, 1975).
9. W. D. Cornell, P. Cieplak, C.I. Bayly, I.R. Gould, K.M. Merz, D.M. Ferguson, D.C. Spellmeyer, T. Fox, J.W. Caldwell and P.A. Kollman. J. Am. Chem. Soc. 117, 5179 (1995).
10. S. J. Plimpton, J. Comp. Phys. 117, 1 (1995).
11. W. Smith and T. R. Forester, DL-POLY, Daresbury, Nr. Warrington, England: The Council for the Central Laboratory of the Research Councils, 1996.
12. High Performance Computational Chemistry Group. "NWChem, A Computational Chemistry Package for Parallel Computers, Version 4.5" Pacific Northwest National Laboratory, Richland, WA, 2003.
13. T. Mori and K. Tanaka, Acta Metallurgica 21, 571 (1973).
14. Y. Benveniste, Mech. Mat. 6, 147 (1987).
15. G. M. Odegard, S. J. V. Frankland, and T. S. Gates, "The Effect of Chemical Functionalization on Mechanical Properties of Nanotube/Polymer Composites", 44th AIAA/ASME/ASCE/AHS/ASC Structures, Structural Dynamics, and Materials Conference, AIAA-2003-1701.
16. R.M. Christensen, Mechanics of Composite Materials (John Wiley & Sons, New York, 1979). p. 95.
17. J. G. Berryman, J. Acoust. Soc. Am. 68, 1820 (1980).
18 . S. Wolfram, The Mathematica Book, 4th Ed, (Wolfram Media, Cambridge, MA, 1999).
19. D. H. Kaeble, J. Moacanin, and A. Gupta "Physical and Mechanical Properties of Cured Resins", in Epoxy Resins: Chemistry and Technology, 2nd edition, C. A. May, ed. (NewYork: Marcel Dekker, Inc., 1988) pp. 603-651.

Mater. Res. Soc. Symp. Proc. Vol. 887 © 2006 Materials Research Society   0887-Q03-03

Effect of Nanoparticles on the Thermal Stability of Polymers

Juan González-Irun, Ana Garcia[1], Ramón Artiaga[2], Luís Liz-Marzán[1], David Hui, Mircea Chipara[3]

Department of Mechanical Engineering, University of New Orleans, LA 70148, USA.
[1]Departamento de Química Física, Universidade de Vigo, Vigo, Spain
[2]Departamento de Ingeniería Industrial II, Universidade da Coruña, Ferrol, Spain
[3]Department of Chemistry, Indiana University, USA.

ABSTRACT

The thermal stability of polymethylmethacrylate loaded with silica nnaoparticles is investigated. The temperature dependence of the mass of of polymethylmethacrylate-silica nanocomposites has been fitted accurately with a single Boltzmann-like dependence. The effect of surface functionalization, silica nanoparticles size, and silica nanoparticles concentration is reported. It was observed that the thermal stability of polymethylmethacrylate increases as the polymer matrix is filled with nanoparticles. Surface functionalization produced a modest increase in the thermal stability (compared to composites with silica nanoparticles without surface modification).

1. INTRODUCTION

The use of nanoparticles as filler-reinforcement of polymer matrix is of high interest since an increase of the mechanical properties of polymeric matrices was reported for low concentrations of nanotubes. The thermal properties of the polymeric matrix (glass transition temperature, melting temperature, and crystallization temperature) are affected by the interactions between nanoparticles and macromolecular chains. For many industrial applications, where the material may be subjected to high temperatures, it is also important to know the effect of nanoparticles on the thermal stability of the polymeric matrices. It is generally accepted that an increase of the inorganic filler content results in higher thermal stability. Nevertheless, it has been also reported that at certain temperatures the nanocomposite may show higher degradation rate than the pristine polymer. The effect of nanoparticle content and nanoparticle features on the thermal stability of different polymers has been investigated by dynamic thermogravimetrical analysis. Silica nanospheres with narrow particle size distribution were dispersed in polymethyl methacrylate. The experiments were carried out both in air and inert atmosphere with pristine and filled polymers. Finally, the plots were fitted to a degradation model in order to quantify mare accurately the degradation processes.

2. EXPERIMENTAL

The synthesis of colloidal silica spheres was performed by Stöber method [1]. Particles of 350 nm (diameter) have been subjected to surface modification that includes two different coatings. The first coating was performed by addition of 3-(trimethoxysilyl) propyl methacrylate[3] (TPM) in excess, followed by slow distillation of ethanol and final cleaning by repeated centrifugation. For the second coating octadecyltrimethoxy silane[4] (OTMS) addition was performed by adding 10% OTMS chloroform solution in the dispersion of particles in ethanol under the presence of ammonia. Particles without any surface modification were used as reference.

Final samples were prepared by dissolving the photoinitiator DMPA (2,2 dimethoxy 2 phenylacetone) in the silica-MMA dispersion, followed by injection in 20x5x1mm glass-cells made by gluing two glass slides spaced by a Teflon with adhesive tape. The cell was irradiated under a UV-lamp for 6 h to achieve complete polymerization of PMMA. The experiments were carried out in a Rheometric STA 1500+. Samples of about 12 mg were placed in open aluminium crucibles. A constant heating rate was applied from room temperature to 600 °C. Some experiments were conducted using a purge of 50 ml/min of dry air. In other cases, 50 ml/min of N_2 were used during the heating ramp. Then the gas was changed to air and the temperature was kept isothermal for 20 minutes. The samples consisted of pristine polymers and silica nanocomposites with different filler contents and different particle size.

3. RESULTS AND DISCUSSIONS

The thermal decomposition of all these samples is well described by a Boltzmann like dependence, with no extreme points and a single inflection point, described by the expression:

$$m(T) = \frac{A[\exp - \alpha(T - T_0)] + m_\infty}{1 + C[\exp - \alpha(T - T_0)]} \qquad (1)$$

Where m is the mass of the sample at the temperature T, m_x is the mass of the sample at vary large temperatures, T_0 is a temperature close to the inflection temperature of the sample mass dependence on temperature, α describes the kinetics of the thermal degradation process, and C is a constant. This model predicts that the mass of the sample at T_0 is:

$$m(T_0) = \frac{A + m_\infty}{1 + C} \qquad (2)$$

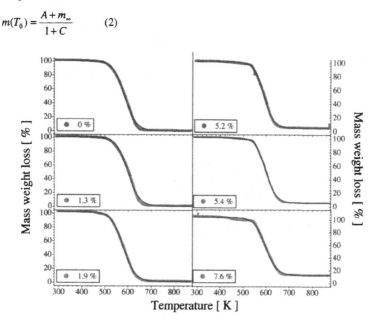

Figure 1. The temperature dependence of the mass of some PMMA based samples.

As it is observed from Figure 1, expression (1) describes accurately the temperature dependence of the mass for pristine PMMA and PMMA doped with silica nanoparticles (with and without surface treatment). The thin black line represents the best fit and the wide gray circles the experimental data.

Figure 2 shows the effect of nanometer-sized silica particles on the thermal stability of PMMA. It is observed that the loading of PMMA with silica nanoparticles produces a weak increase in the thermal stability of the polymeric matrix. From Figure 2 it is observed that the thermal stability of the polymeric matrix is enhanced as the concentration of nanometer sized silica particles is increased (up to about 10 %).

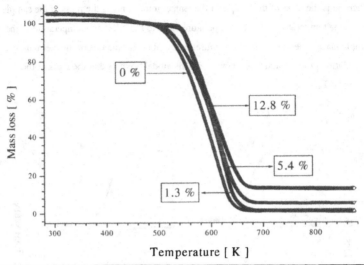

Figure 3. TGA traces obtained in air from PMMA samples with different contents of 350 nm silica nanoparticles without any surface treatment.

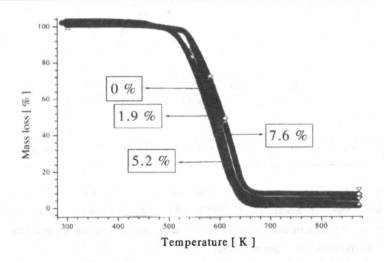

Figure 2. TGA traces obtained in air from PMMA samples with different contents of 350 nm silica nanoparticles with TPM surface treatment.

Figures 3 and 4 depict the effect of silica nanoparticles (after a surface treatment with TPM and OTMS respectively) on the thermal degradation of PMMA. As observed by comparing Figure with Figures 3 and 4, the effect of surface treatment on the thermal stability of PMMA is rather modest. However, best results were obtained with the OTMS surface treatment.

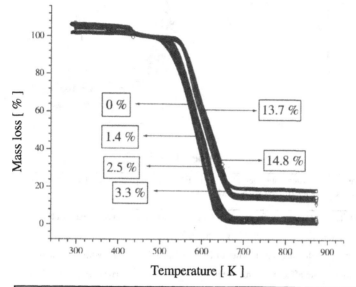

Figure 4. TGA traces obtained in air from PMMA samples with different contents of 350 nm silica nanoparticles with OTMS surface treatment.

EFFECT OF PARTICLE SIZE

Figure 5 shows that the effect of particle size is small. A weak tendency to increase the thermal stability of PMMA as the size of silica particles was increased has been observed.

CONCLUSIONS

TGA investigations on the thermal stability of pristine PMMA, PMMA loaded with various concentrations on silica nanoparticles, and surface treated silica nanoparticles dispersed in PMMA have been reported. In most cases the temperature dependence of the mass loss was accurately described by a single Boltzmann-like dependence. A weak Boltzmann like component has been noticed at lower temperatures in the PMMA

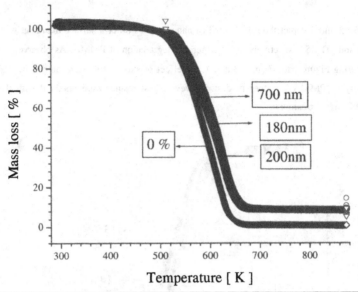

Figure 5. TGA traces obtained in air from PMMA samples with different sizes of silica nanoparticles treated with TPM.

composites containing surface treated silica nanoparticles (OTMS treatment) at a concentration of silica particles larger than 10%.

An increase of the thermal stability of PMMA upon the loading with silica nanoparticles was noticed. The effect was only weakly enhanced by the surface treatment of silica nanoparticles.

Further experiments are required to understand the effect of nanoparticles size on the thermal stability of polymers. The as obtained data seem to indicate that the stability of PMMA is increased as the diameter of silica particles (subjected to the TPM surface treatment) is increased. This is the variance with the hypothesis that the increase of the thermal stability of the polymer is controlled by surface processes as smaller particles correspond to larger surfaces, for the same volume concentration of nanoparticles.

ACKNOWLEDGEMENTS

This research has been funded by US ARMY STTR #A2-1299 grant awarded to the Chemistry Department of Indiana University (Bloomington) and to PartTec Ltd. And by ONR funding (University of New Orleans). Financial support from the Spanish Xunta de Galicia (Project no. PGIDIT03TMT30101PR) and Ministerio de Educación y Ciencia (Grant MTM2005-00429, ERDF included) is gratefully acknowledged.

REFERENCES

1. W. Stober, A. Fink, E. Bonh, J. Coll. Interf. Sci. 26 (1968) 62
2. G. H. Bogush, M. A. Tracy, C. F. Zukoski.. Journal of Non-Crystalline solids 104 (1988) 95-106
3. A. P. Philipse, A. Vrij, J. Coll. Interf. Sci 128 (1989) 121-136
4. W.Wang, B. Gu, L. Liang, W. Hamilton, J. Phys. Chem.B. 107 (2003) 3400-3404

Mater. Res. Soc. Symp. Proc. Vol. 887 © 2006 Materials Research Society   0887-Q02-01

Fire Retardancy and Morphology of Nylon 6-clay Nanocomposite Compositions.

Kadhiravan Shanmuganathan [1], Sandeep Razdan[2], Nick Dembsey[3] Qinguo Fan[1], Yong K Kim[1], Paul D Calvert[1], Steven B Warner[1] and Prabir K Patra[1].
1. Department of Materials and Textiles, University of Massachusetts - Dartmouth, North Dartmouth, MA-02747.
2. Department of Material Science and Engineering, Rensselaer Polytechnic Institute, Troy, NY.
3. Dept of Fire Protection Engineering, Worcester Polytechnic Institute, MA-01609

ABSTRACT

We investigated the effect of organically modified clay on the thermal and flammability behavior of nylon 6 nanocomposites. We also used zinc borate along with layered silicate with an aim of achieving synergistic effect in flame retardancy. It is found that addition of 10 wt% clay reduced the onset decomposition (5% wt loss) temperature of nylon 6 by 20°C, while addition of 5 wt% zinc borate and 5 wt% clay in combination reduced it by around 10°C. Differential thermogravimetric analysis indicated that the peak decomposition temperature was not affected by the addition of clay, but the rate of weight loss decreased with increasing clay concentration. The horizontal burning behavior of the nanocomposite films of approximately 0.5mm thickness changed with additive concentration. The nanocomposites with 2.5 wt% and 5 wt% clay burned for almost the same duration as neat nylon 6 but dripping was reduced. The 10 wt% clay nanocomposite sample burned without any dripping and the flame spread rate was reduced by 25-30%. The burn rate of 5 wt % zinc borate/5 wt% clay nanocomposite sample was about 20% higher than that of 10 wt% clay nanocomposite sample, which could be attributed to varying char morphology. Scanning electron microscopy images of the 10wt% clay nanocomposite char surface and cross- section revealed an integrated layer of clay platelets with increasing density gradient from the center to the surface, while the 5 wt% zinc borate/5 wt% clay nanocomposite char appeared foamy and porous. The 5 wt% zinc borate and 5 wt% clay sample developed into a very good intumescent system in cone calorimeter test, swelling about 10-13mm height prior to ignition forming a cellular char structure. This was as effective as the 10wt% clay nanocomposite sample in reducing the heat release and mass loss rate of nylon 6 by around 65%. Fourier transform infrared spectroscopy of the 10 wt% clay nanocomposite char showed the presence of amides, indicating possible residual polymer within the shielded char.

INTRODUCTION

Nanocomposites are a distinct class of composite materials showing impressive performance for multifunctional applications at a very low filler level of 2-5 wt%. They are defined by the particle size of the dispersed phase having at least one dimension less than 100 nm [1]. Polymer-layered silicate nanocomposites (PLSN) have been in the realm of research since 1961 [2] owing to their enhanced mechanical [3], thermal [4] and barrier properties [5] over conventional composites with micron size fillers. The enhancement in properties is largely attributed to the aspect ratio and intercalation ability of the layered silicates, resulting in high specific surface area and hence more potential for polymer filler interaction [6]. Nanocomposites show great promise as effective fire retardants for many polymers [7-12]. Because of the upcoming restrictions on some of the current popular flame retardants, the flame retardancy of nanoparticle filled systems or any non-halogen alternatives need to be improved to meet the new standards. As a result interest in this area of research has recently increased. Though the significant reduction in heat release rate of nanocomposites was reported for quite some time [2], the flame retardancy of polymer filled with nanoparticles such as clay, carbon nanotubes or silica has not further improved to achieve industrial acceptance. Also the effect of nano fillers on thermal stability and flame retardancy of nanocomposite fibers is not clear.

Work done in the previous years by our group and others were focused on understanding the mechanism behind the flame retardancy of polymer layered silicate nanocomposites. It has been found that the addition of 2-5 wt% of montmorillonite reduces the peak heat release rate by 50-60 % compared to neat nylon 6 as found by cone calorimeter studies on slabs of 4-5 mm thickness. The formation of a surface char which acts as a thermal insulation and mass barrier reduces the mass loss rate and hence the rate of heat release [13, 14]. The heat of combustion, CO_2 and CO yields are not changed significantly, which indicates the predominant role of layered silicates in the condensed phase rather than the gas phase. Since the flame retardant effects of layered silicates is more physical than chemical the sample geometry would strongly influence the burning behavior of nanocomposite films. The nanocomposite films of thickness less than 1mm or a fabric of nanocomposite fibers would behave much different when subjected to a flame. As the sample becomes thinner, the concentration of layered silicates would become critical in providing a significant effect. In that case we need to either increase the concentration of layered silicates to a level sufficient to form a network of platelets and hence a strong mass barrier or alternatively the flame retardant effect of layered silicates has to be augmented by another flame retardant additive. We studied the effect of clay concentration on the flame retardant behavior of nanocomposites in relation to polymer layered silicate interactions and char morphology. We also investigated the effect of zinc borate in enhancing the flame retardancy of polymer-layered silicate nanocomposites.

We chose zinc borate in order to reinforce the flame retardant effect of layered silicates through the formation of a stable char. Decomposition of zinc borate releases B_2O_3 moiety, a low melting glass that can stabilize the char. It also releases its water of hydration (about 13-15%) at 290-450 °C which can blow the char to a foamy structure [15]. Thus it could significantly influence the char morphology and hence the flame retardancy of polymer layered silicate nanocomposites. Also it is thermally stable in the processing temperature of nylon 6.

EXPERIMENTAL

Materials

Nylon 6 was supplied by Honeywell Inc and montmorillonite (1.34 TCN), modified with methyl dihydroxyethyl hydrogenated tallow ammonium was provided by Nanocor Inc. Hence the term layered silicate or clay used in our literature refers to organically modified montmorillonite (OMMT). The specific surface area of OMMT is around 750 m^2/g and the cation exchange capacity is 92 meq/100g (as per manufacturer). Nylon 6 pellets were dried in a vacuum chamber at 80°C for 16 hours and OMMT was dried for 4 hours at 100°C prior to melt mixing. Nanocomposite hybrids with 2.5 wt% (NCH 2.5), 5 wt % (NCH 5), and 10 wt% (NCH 10) of OMMT were prepared by melt compounding using Brabender twin blade roller mixer. Also, nanocomposite with 5 wt% OMMT and 5 wt % zinc borate (NCH OMMT/ZnB) and nanocomposite with 5wt% ZnB were prepared. Melt compounding was done at 240°C for 10 minutes at 90 rpm. The samples were pressed into films of approx 0.5 mm thickness using Carver hot press under a pressure of 10000 psi at 240°C for 2 min.

Characterization

Thermal Analysis

The pyrolysis behavior of nanocomposites and neat samples was analyzed using TA instruments model Q500 Thermo gravimetric analyzer (TGA). The samples were heated in a nitrogen atmosphere from room temperature to 600°C at a heating rate of 20°C/min. The kinetics of the decomposition process was studied by differential thermo gravimetric analysis and the activation energy of the decomposition process was determined assuming a first order reaction for polymer pyrolysis using the Broido equation [16],

$$\ln[\ln(\frac{1}{x})] = \frac{-E_a}{RT} + \text{cons tan t}$$

(1)

where E_a = activation energy for the reaction,
T = absolute temperature in Kelvin
R = 8.314 J/mole °K.

$$x = \frac{W_t - W_\infty}{W_0 - W_\infty}$$

(2)

where W_t is the weight of the sample at any time t, W_0 and W_∞ are initial and final weight of sample respectively.

Flammability testing

The nanocomposites films of approx 0.5mm thickness were subjected to flame testing (FMVSS 302) using an Atlas horizontal flame spread tester. The flame height was set to about 1.5 inch and the ignition source was applied for 15 sec.

Flammability testing was also done on selected samples using cone calorimeter. Nanocomposite slabs of 100 mm X 100 mm and about 3.0±0.1 mm thickness were subjected to an incident heat flux of 35 kW/m^2 and the ignition and heat release characteristics were observed.

Char morphology was imaged using scanning electron microscope (JEOL 2000) with an accelerating voltage of 15 kV. Also Fourier transform infrared spectroscopy was done on the char using Digilab Excalibur series FTS 3000 Max in diffuse reflectance mode.

RESULTS AND DISCUSSIONS

Thermal Behavior

Sample	5% wt loss temp (°C)	10% wt loss temp (°C)	Peak decomposition temp (°C)	Peak decomposition rate (%/min)	Residue (%)	Activation energy X 10^5 (J/mol)
Neat nylon 6	432	447	488	45.59	1.009	2.24
NCH 2.5	427	444	486	45.87	2.242	2.25
NCH 5	422	440	490	37.10	5.506	1.85
NCH 10	412	436	487	39.24	8.769	1.59
NCH Zn B	420	438	484	41.36	11.03	1.82

Table I: TGA and DTGA data of nylon 6 and nanocomposites

The onset decomposition temperatures (5% and 10% wt loss temperature), peak decomposition temperature, peak decomposition rate, % residue, activation energy of decomposition of nanocomposites and neat samples are given in Table I. The values reported are average of 3 tests The TGA graphs and data showed reduction in the onset decomposition temperature of nanocomposite as compared to neat nylon 6 (Figure 1a) and the difference increased with increasing concentration of OMMT. However, the peak decomposition temperature was almost same and decomposition rate reduced with increasing concentration of OMMT (Figure 1b). The amount of additional carbonaceous residue was negligible accounting for the OMMT content.

The influence of layered silicates (OMMT) on the decomposition behavior of nanocomposites can be considered in dual aspects. The presence of enhanced polymer layered silicate interactions favor the thermal stability of the polymer. On the other hand the organic modifier present in the montmorillonite decomposes between 200-300°C and causes significant polymer matrix degradation during the melt processing of polymer organoclay mixtures (17). It is found that OMMT and water combine to catalyze the degradation of polymer (18). Hence, the decomposition behavior of nanocomposites would be influenced by the concentration of OMMT. From the values of decomposition temperature it is difficult to confirm the degradation effects of OMMT as a function of concentration. Therefore the estimated weight loss of nanocomposites was compared with the actual weight loss as a function of temperature (Table II). The estimated weight loss was calculated using the formula,

$$M_c(T) = W_p * M_p(T) + W_f * M_f(T)$$
$$\tag{3}$$

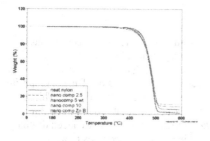
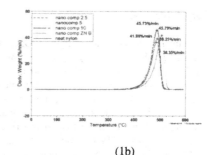

(1a)   (1b)

Figure 1: (a) TGA thermograms of nylon 6 and nanocomposites (b) DTGA thermograms of nylon 6 and nanocomposites.

Table II: Weight loss data of nylon 6 and nanocomposites

Temp °C	NCH 2.5 M_c (%)		NCH 5 M_c (%)		NCH 10 M_c (%)	
	Estimated	Observed	Estimated	Observed	Estimated	Observed
300	99.27	99.30	98.98	99.16	98.40	99.46
350	98.99	99.03	98.48	98.79	97.59	98.74
400	98.06	98.15	97.52	97.64	96.40	96.13
450	88.24	86.59	87.9	84.70	87.20	83.76

where $M_c(T)$, $M_p(T)$ and $M_f(T)$ are respectively the percentage mass of nanocomposite, nylon 6 and OMMT not yet decomposed at a particular temperature T, and W_p and W_f are the weight fractions of nylon 6 and OMMT in the nanocomposite.

It is clear from the table that up to 400°C, the observed % fraction of nanocomposite not yet decomposed is higher than the estimated values based on weight fraction of polymer and OMMT. Above that temperature the observed values are less than the estimated values and the difference increases with increase in concentration of clay. This clearly shows the polymer - clay interactions and degradation effects of clay working against each other on the decomposition behavior of nylon 6. The values of activation energy for the decomposition process also decreased with increasing concentration of clay. However the values of decomposition temperature and activation energy for 5 wt % Zinc borate/5 wt% clay nanocomposite was higher than 10 wt% clay nanocomposite and almost close to 5 wt % clay nanocomposite showing the dominance of degradation effect of clay at higher concentrations.

In the case of nanocomposites with a lower clay loading (upto 5 wt %), which has partially exfoliated /intercalated structure, the polymer layered silicate interaction is enhanced which compensates the degradation effects in the polymer. With a 10 wt% loading the clay particles are not sufficiently intercalated and there is less interaction between the polymer and layered silicates. This causes the degradation effect to dominate and result in lower onset decomposition temperature. Nevertheless the peak decomposition temperature is almost same and the decomposition rate decreases with increasing concentration of OMMT.

Flammability Behavior

The flame spread behavior of nylon 6/OMMT polymer films with different concentrations of OMMT (2.5, 5 and 10wt %) is given in table III where the reported values are average of 3 tests. The 2.5 wt% and 5 wt % OMMT nanocomposite films burned much the same way as neat nylon films, though the dripping tendency was reduced. However the 10 wt% OMMT nanocomposite film burned without any dripping and the flame spread rate was reduced by 20-30%. This assumes significant importance in practical applications. Dripping initiates flame spread to surrounding materials and also increases risk of injuries during fire hazards. Reduced dripping or no dripping would be beneficial from safety point of view. The nanocomposite with 5wt% OMMT and 5wt% zinc borate showed very less dripping. However the flame spreading rate was higher than 10wt% OMMT nanocomposite film (Table III).

Table III: Burning behavior of nylon 6 and nanocomposites

Sample	Burning behavior	Flame spreading rate (inch/min)
Neat nylon	Pronounced dripping	1.803
2.5 wt% OMMT nanocomposite	Drips	1.723
5 wt% OMMT nanocomposite	Reduced dripping	1.380
10 wt% OMMT nanocomposite	No dripping	1.222
5wt% OMMT/ 5wt% Zn B nanocomposite	Very less dripping	1.625

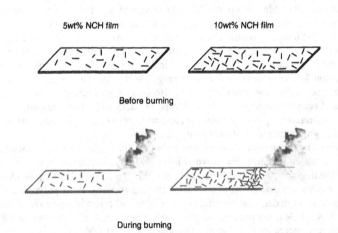

5wt% NCH film   10wt% NCH film

Before burning

During burning

Figure 2: Schematic representation of horizontal burning behavior of films.

When the concentration of MMT is high enough to form a network structure and the material doesn't drip during burning, then the flame has to advance through the burnt clay. With gasification of the polymer near the flame front, the concentration of OMMT becomes high enough to form a significant mass barrier hindering the flow of volatile gases. This would significantly reduce the mass loss rate and heat release rate. This phenomenon can be schematically represented as in figure 2.

Morphology of Char

The microstructure of the char surface of the burnt nanocomposite films observed using scanning electron microscopy seems to explain well the difference in burning behavior of the nanocomposite films (Figure 3).

The neat nylon and the NCH 2.5 film did not form an interconnected char and dripped pronouncedly during burning. The NCH 5 and NCH 10 films formed significant char on burning. However a clear difference in char morphology could be observed between the NCH 5 and NCH 10 films. The NCH 5 film produced significant char but the char surface revealed lot of cracks and pores (figure 3a). However the NCH 10 film with 10 wt % OMMT formed a consolidated char (figure 3b) with very few cracks or pores. SEM image of the char cross section revealed densely accumulated clay platelets on both surfaces of char with density gradient from the center to the surface (figure 3c). The high aspect ratio and the platelet geometry of the OMMT led to a well interconnected network of platelets forming a protective shield. This interconnected network of platelets and viscosity effects could have restricted the dripping of the 10 wt % OMMT nanocomposite film during burning. Also this could result in significant mass barrier effect compared to NCH 5 film, which could explain the difference in burning rate of the films. The addition of 5wt% zinc borate along with 5wt% OMMT instead of 10wt% OMMT resulted in a significant change in char morphology (figure 3d). The char appeared to be foamy which might be due to water released during the decomposition of zinc borate. Most zinc borates in commercial use are hydrates $(xZnO.yB_2O_3.nH_2O)$ which release their water of hydration (about 13-15%) at 290-450°C. In addition to absorption of heat and dilution of fuel the release of water serves to blow the char to foam [9]. The difference in char morphology could have a significant effect on the thermal insulation and barrier properties. However cone calorimeter studies would be required to obtain more information and correlate char morphology and burning behavior.

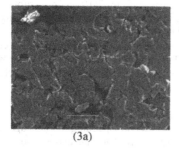

(3a)

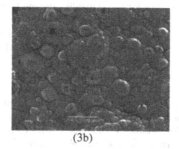

(3b)

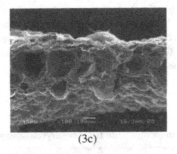

(3c)   (3d)

Figure 3: SEM micrographs of (a) NCH 5 char (b) NCH 10 char (c) NCH 10 char cross section (d) NCH ZnB char

Cone Calorimeter Studies

The cone calorimeter is being used extensively for flammability testing nowadays as the test reveals a lot of data useful for interpreting the combustion behavior of materials in real fires [2]. The burning behavior of nanocomposite slabs is shown in figure 4. It is found that the burning behavior of 10 wt% OMMT nanocomposite slab differs much from the nanocomposite with 5wt% OMMT and 5wt% zinc borate. In the 10 wt% OMMT nanocomposite the specimen surface started to granulate before ignition and ignition flash occurred prior to full ignition. After ignition, specimen surface burned evenly across entire surface and a well consolidated layer of char was obtained (figure 4a and 4b). However in the case of nanocomposite with 5wt% OMMT and 5wt% zinc borate, the specimen surface started to bubble before ignition and intumesced approximately 5mm before ignition. After ignition, specimen surface burned unevenly across surface for several seconds before burning occurred over entire surface. Also the specimen intumesced significantly to about 10-13mm almost hitting the spark igniter. At the end a cellular char structure of about 10mm in height was obtained as shown in figure 4c and 4d. In order to distinguish the role of zinc borate and OMMT on char formation process one sample with nylon 6 and 5wt% zinc borate was also studied. Here the specimen surface began to bubble and intumesce in a localized manner and ignition occurred unevenly for 7-10 seconds before flames were present over (figure 4e and 4f). Some fragments of char were observed after burning instead of a consolidated layer. Thus it was believed that the release of water during the decomposition of zinc borate leads to bubbling and intumescence. However in the absence of char forming agent the intumescence could not develop into a well blown char structure and affect flame retardancy. The montmorillonite platelets forming a network and covering the surface serves a scaffold for forming a stable and cellular char structure along with the intumescent action of zinc borate which could not be obtained with zinc borate alone.

The heat release rate, mass loss rate, heat of combustion and smoke release data are presented in table IV and table V with values in parentheses corresponding to % change. The average values for each sample were recalculated after truncating the data corresponding to the tail portions of the respective heat release and mass loss curves. This is to ensure that signals and noise in the beginning due to flashes and those corresponding to burning of samples in the edges are eliminated and the values represent the burning behavior of samples in the stable burning regime.

The peak heat release rates (HRR) of the nanocomposites were significantly lower than that of nylon 6. The addition of 10wt% OMMT reduced the peak HRR by about 67% from 948 kW/m^2 to 310 kW/m^2 whereas addition of 5wt % OMMT and 5wt% ZnB reduced the peak HRR by 69%. This is significantly higher than the reduction obtained with 5wt% OMMT (54%) and 5wt% ZnB (34%) alone.

The average HRR values which are more consistent and meaningful for comparing the burning behavior of the samples also concur well with peak HRR data and revealed almost similar percentage reduction. The average mass loss rate was also reduced by more than 60% by the addition of 10wt% OMMT or 5wt% OMMT and 5wt% ZnB which mirrored the trend observed in HRR data. On the other hand the effective heat of combustion was almost same for all samples except the 10wt% OMMT nanocomposite sample showing slightly lesser heat of combustion. The heat release rate is a function of mass loss rate and heat of combustion. With the heat of combustion remaining same it is the significant change in the rate of mass loss that has resulted in a drastic reduction in heat release rate.

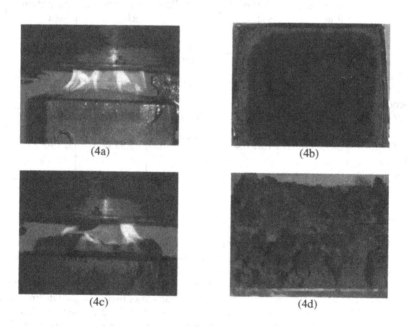

(4a)

(4b)

(4c)

(4d)

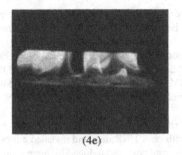
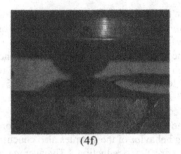

(4e)　　　　　　　　　　　　　　　(4f)

Figure 4: Digital images of burning behavior and char formation (a & b) NCH 10, (c & d) NCH ZnB, (e & f) Nylon6/5wt% ZnB composite.

Table IV: Heat release and mass loss data of nylon 6 and nanocomposites

Sample	Peak Heat Release Rate (kW/m^2)	Avg Heat Release Rate (kW)	Avg Mass Loss Rate (g/s)	Avg Heat of Combustion (kJ/g)	Sp. Extinction Area (m^2/g)
Nylon 6	948	5.04	0.18	28.04	0.099
5wt% OMMT nanocomposite	433 (54)	2.58 (49)	0.09 (50)	28.62	0.177
10wt% OMMT nanocomposite	310.01 (67)	1.79 (65)	0.07 (61)	25.77 (8)	0.182
5wt%ZnB nanocomposite	624.80 (34)	2.92 (42)	0.10 (44)	27.82 (0.8)	0.184
5wt% OMMT/ 5wt%ZnB nanocomposite	289.25 (69)	1.81 (64)	0.06 (66)	27.37 (2.3)	0.196

<u>**Char analysis**</u>

The SEM image of the nanocomposite char led us to anticipate entrapment of polymer fragments between the densely accumulated clay platelets on both surfaces of char. Fourier transform infrared spectroscopy (FTIR) was conducted on clay, NCH 5 and NCH 10 char obtained from the flammability tests. In addition to Si peaks at 3670 cm^{-1} and 1031 cm^{-1}, which represents stretching vibration of Si-OH and Si-O, the NCH 10 char shows two additional distinct peaks at 1630 cm^{-1} and 1520 cm^{-1} which were not observed in clay and NCH 5 as shown in figure 5. The peaks at 1630 cm^{-1} and 1520 cm^{-1} could be assigned to amide I and amide II respectively [19] which might indicate possible unburnt polymer fragments or recession within the shielded

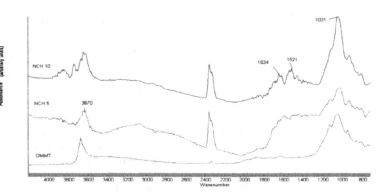

Figure 5: FTIR spectra of nanocomposite char

char. At a higher concentration, the OMMT platelets collapse to form a well integrated shield in the case of NCH 10 char as compared to NCH 5 char (SEM micrograph, figure 3). This might have caused entrapment of polymer fragments and retention of unburnt chain fragments within the shielded char. These spectra are obtained from burnt char of nanocomposite film of thickness of 0.5mm. With increase in film thickness the shielding effect of the char would be more pronounced and could result in significant retention of polymer. This would reduce the amount of combustible polymer available for burning and might influence the burning behavior strongly.

CONCLUSIONS

We investigated the effect of layered silicate concentration and combined effect of zinc borate and layered silicates on the thermal stability and flame resistance of Nylon6/layered silicate nanocomposites. We focused our investigation on the effectiveness of the char formation process on the flame spread behavior of thin nanocomposite films of approximately 0.5 mm thickness. The onset of decomposition (5% wt loss temperature) decreased with increasing concentration of OMMT. The addition of 5 wt% zinc borate and 5 wt% OMMT in combination minimized the degradation effect of nylon6 as compared to the addition of 10 wt% OMMT. However the peak decomposition temperature was almost same for all nanocomposites but the rate of weight loss decreased with increase in OMMT concentration. It is found that in thin films, with increase in concentration of OMMT from 5 wt% to 10 wt%, the dripping of polymer is completely suppressed and the nanocomposite film remains intact during burning. While the nanocomposites with 2.5 wt% and 5 wt% clay burned for almost the same time as neat nylon 6 with reduced dripping tendency, the 10 wt% clay nanocomposite sample burned without dripping and the flame spreading rate was reduced by 25-30%. Addition of 5 wt% zinc borate and 5 wt% clay restricted the dripping of nylon 6 but not completely and also the flame spreading rate was slightly higher than that of 10 wt% nanocomposite sample. SEM of the burnt char showed a completely different char morphology for the two samples which could be the reason behind the difference in burning behavior. This demonstrates the effectiveness of the high aspect ratio OMMT platelets in forming a significant mass barrier and affecting the burning behavior in the case of thin films. The reinforcing effect of zinc borate on char formation of Nylon 6/OMMT system was however evident in cone calorimeter studies. The 5 wt% zinc borate and 5 wt% clay sample intumesced significantly, swelling about 10-13mm height prior to ignition and resulted in a cellular char structure after burning. This char structure was not observed either with OMMT or zinc borate alone and resulted in a significant reduction in heat release and mass loss rate of nylon 6 (about 65%) comparable to 10 wt% OMMT nanocomposite sample. It could be concluded that though decomposition of zinc borate releases water leading to bubbling and intumescence, a char forming agent is needed to develop the intumescence into a well blown char structure and affect flame retardancy. Thus the zinc borate/OMMT system could be an effective intumescent flame retardant for nylon 6 in form of slabs. Fourier transform infrared spectroscopy (FTIR) of the 10 wt% OMMT nanocomposite char showed the presence of amides which might indicate possible unburnt polymer or recession within the shielded char. This was not observed in other chars. Thus 10wt% OMMT is sufficient enough to restrict the dripping of polymer and form a protective shield like char even in thin films of about 0.5mm thickness. Further enhancement of flame retardancy could be achieved by combining the physical effect of OMMT with chemical action either through coating or intercalating layered silicates with flame retardant additives.

ACKNOWLEDGEMENTS

We thank Mr.Leo Barish and Mithun Shah for their help and discussions with SEM studies, Randall Harris (WPI) and Frank Kozekowich (Quaker fabrics) for helping with flammability tests. We also thank Nanocor Inc for providing the layered silicates and Honeywell Inc for providing Nylon 6. Our special thanks are due to National Textile Center for funding this study through US Department of Commerce grant 02-0740 under the project M02-MD08.

REFERENCES:

1. E. P. Giannelis, Advanced Materials. **8**, 29 (1996)
2. J.W.Gilman, Applied Clay Science. **15**, 31 (1999)
3. A. Usuki, Y. Kojima, M. Kawasumi, A. Okada, Y. Fukushima, T. Kurauchi and O. Kamigaito, Journal of Polymer Science part A; Polymer Chemistry.**31**, 1755 (1993)
4. D.M. Lincoln, R.A. Vaia, Z. Wang, B.S. Hsiao and R. Krishnamoorti, Polymer. **42**, 9975 (2001).
5. K. Yano, A. Usuki and A. Okada, Journal of Polymer Science A. **35**, 2289 (1997).
6. A.R. Horrocks and Price D, editors, Fire Retardant Materials, (Woodhead publishing Limited, Cambridge, England 2000) chap 4, p.204-219.
7. C. Zhao, H. Qin, F. Gong, M. Feng, S. Zhang and M. Yang, Polymer Degradation and Stability. **87**, 183 (2005).
8. S. Su, D.D Jiang and C.A. Wilkie, Polymer Degradation and Stability. **83**, 321 (2004)
9. T. Kashiwagi, R.H Harris Jr, X. Zhang, R.M. Briber, B.H. Cipriano, S.R. Raghavan, W.H. Awad and J.R. Shields, Polymer. **45**, 881 (2004)
10. A.B.Morgan, T. Kashiwagi, R.H. Harris Jr, L.J Chyall and J.W. Gilman, Fire and Materials. **26**, 247 (2002)
11. L. Song, Y. Hu, Y.Tang, R. Zhang, Z. Chen and W. Fan, Polymer Degradation and Stability. **87**,111 (2005)

12. T.H. Chuang, W. Guo, K.C. Cheng, S.W. Chen, H.T. Wang and Y.Y. Yen, Journal of Polymer Research. **11**, 169 (2004).

13. T. Kashiwagi, R.H. Harris Jr, X. Zhang, R.M. Briber, B.H. Cipriano, S.R. Raghavan, W.H. Awad and J.R. Shields, Polymer. **45**, 881 (2004)

14. G. Inan, P.K. Patra, Y.K Kim and S.B.Warner, Mat.Res.Soc.Symp.Proc. **788**, L 8.46 (2003)

15. A.R. Horrocks and Price D, editors, Fire Retardant Materials, (Woodhead publishing Limited, Cambridge, England 2000) chap 2, p. 31-68.

16. A. Broido, Journal of Polymer Science, Part A-2. **7**, 1761 (1969).

17. T.D. Fornes, P.J.Yoon and D.R. Paul. Polymer, **44**, 7545 (2003).

18. R.D.Davis, J.W. Gilman and D.L. VanderHart, Polymer Degradation and Stability. **79**, 111 (2003).

19. G. Chen, D. Shen, M. Feng and M. Yang, Macromolecular Rapid Communications. **25**, 1124 (2004).

Nanometer-Scale Degradation
Processes in Magnetic and
Superconducting Materials

Mater. Res. Soc. Symp. Proc. Vol. 887 © 2006 Materials Research Society 0887-Q11-02

Large magnetoresistance in oxide based ferromagnet / superconductor spin switches

V. Peña[1], N. Nemes[2], C. Visani[1], J. Garcia-Barriocanal[1], F. Bruno[1], D. Arias[1] #), Z. Sefrioui[1], C.Leon[1], S. G. E. Te Velthuis[3], A. Hoffmann[3], M. Garcia-Hernandez[2] and J Santamaría[1]

[1] GFMC, Departamento de Física Aplicada III, Universidad Complutense de Madrid, 28040 Madrid, Spain
[2] Instituto de Ciencia de Materiales de Madrid (ICMM-CSIC). 28049 Cantoblanco. Madrid.
[3] Materials Science Division, Argonne National Laboratory, Argonne, Illinois 60439, USA

ABSTRACT

We report large magnetoresistance (in excess of 1000%) in ferromagnet / superconductor / ferromagnet structures made of $La_{0.7}Ca_{0.3}MnO_3$ and $YBa_2Cu_3O_7$ in the current in plane (CIP) geometry. This magnetoresistance has many of the ingredients of the giant magnetoresistance of metallic superlattices: it is independent on the angle between current and magnetic field, depends on the relative orientation of the magnetization in the ferromagnetic layers, and takes very large values. The origin is enhanced scattering at the F/S interface in the anti parallel configuration of the magnetizations. Furthermore, we examine the dependence of the magnetoresistance effect on the thickness of the superconducting layer, and show that the magnetoresistance dies out for thickness in excess of 30 nm, setting a length scale for the diffusion of spin polarized quasiparticles.

INTRODUCTION

When a superconductor is placed in contact with a ferromagnet both long range phenomena may compete at the interface [1,2] giving rise to a variety of exotic phenomena like π- junctions, spatially modulated order parameter, etc [3,4]. There has been substantial activity in the past directed to study the F/S interplay in heterostructures containing transition metal superconductors (low Tc) and ferromagnets [5-10]. With the (re) discovery of CMR materials there has been renewed activity in the field with heterostructures involving High Tc superconductor (HTS) and colossal magnetoresistance materials (CMR), which incorporate a number of interesting new ingredients [11-16]. The larger *Tc* of the HTS sets an energy scale for the condensation energy which is comparable to the exchange coupling of the ferromagnet. This originates strong competition between both long range phenomena in F/S heterostructures. The short coherence length of the HTS in the c direction (0.1-0.3 nm) enables superconductivity to survive even in very thin layers in the presence of a ferromagnet. Furthermore, the unconventional d-wave symmetry of the SC order parameter may give rise to novel quantum phenomena, related to the occurrence of Andreev bound states at the interface with the ferromagnet. In addition, the high degree

95

of spin polarization of the conduction band of the manganites enhances the FM/SC competition and may open the door to important spin dependent transport effects yielding (useful) magnetoresistance. Finally, it has been demonstrated that many of these perovskite-related oxides can be readily combined, thanks to their good lattice matching and chemical compatibility, yielding heterostructures of exceptionally high quality with very smooth and well-defined interfaces.

In F/S/F structures with a very thin SC layer the Cooper pairs sample different directions of the exchange field within their coherence volume. The proximity induced suppression of the SC order parameter thus will be enhanced when the magnetization in the FM layers is parallel (P) and decreased when it becomes antiparallel (AP), where the effect of opposite exchange fields cancels out. A controlled change of the direction of the magnetization of the FM layers thus enables a modulation of the critical temperature and subsequently introduces novel kinds of magnetoresistance effects which might be extremely large. Very recently, this concept has been tested in conventional NiCu/Nb/NiCu trilayers where a T_c shift of 0.1 K and magnetoresistance values close to 50 % have been obtained [17]. As opposed to the above described positive magnetoresistance effect, in F/S/F structures with highly polarized carriers a negative magnetoresistance effect may occur. This effect is related to the spin accumulation in the SC layers whose thickness is well below the spin diffusion length. The induced spin density within the SC layer (involving states above the SC energy gap) then becomes larger for the antiparallel alignment of the magnetization of the FM layers than for the parallel one. The subsequent T_c suppression due to the related pair-breaking effects can yield to sizeable negative magnetoresistance effects as proposed in Ref. [18]. This spin accumulation picture has been realized in Co/Al/Co planar junctions by Chen and co-workers [19], where the spin diffusion length of Al is known to be very large at low temperatures. Very large magnetoresistance (in excess of 1000%) has been reported in F/S/F structures made of $La_{0.7}Ca_{0.3}MnO_3$ and $YBa_2Cu_3O_7$ [20]. Interestingly the experiment was conducted with the current in the plane of the layers (CIP geometry), which precludes spin accumulation at the interfaces. We propose an explanation in terms of enhanced scattering at the F/S interface in the AP configuration. We also examine the dependence of the magnetoresistance effect on the thickness of the superconducting layer. We show that the magnetoresistance dies out for thickness in excess of 24 nm, setting a length scale for the diffusion of spin polarized quasiparticles.

EXPERIMENT

Samples were grown on (100) oriented $SrTiO_3$ single crystals in a high pressure (3.4 mbar) dc sputtering apparatus at high growth temperature (900 °C). The high oxygen pressure and the high deposition temperature provide a very slow (1 nm / min) and highly thermalized growth which allows the control of the deposition rate down to the unit cell limit. For this study we grew F/S/F trilayers keeping the thickness of the LCMO fixed at 40 unit cells (15 nm) and changed the thickness of the YBCO between 1 unit cell (1.2 nm) and 40 unit cells (48 nm). Structure was analyzed using x-ray diffraction and transmission electron microscopy. Further details about growth and structure can be found elsewhere

[21-23]. X-ray refinement technique using the SUPREX 9.0 software were used to obtain quantitative information about the interface roughness [24]. Tc was measured from 4 point contacts resistance curves as the zero resistance temperature. Magnetization was measured in a SQUID (Quantum design) magnetometer.

RESULTS AND DISCUSSION

We have measured the magnetoresistance with the magnetic field applied parallel to the layers. Current contacts were in the plane of the layers (current in plane geometry) and aligned perpendicular to the magnetic field direction. Field was swept between 1 T and -1T at temperatures fixed along the superconducting resistive transition Figure 1(a) shows R(H) loops at various temperatures for a trilayer sample with 15 unit cells thick YBCO layer.

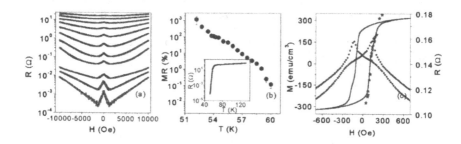

Figure 1. (a). Resistance as a function of magnetic field, R(H) loops, of a F/S/F trilayer [LCMO (40 u.c.) / YBCO (15 u.c.) / LCMO (40 u.c.)] at different temperatures along the resistive transition. Magnetic field, applied parallel to the layers, was swept between -1 T and 1 T fields in an hysteresis loop sequence. Temperatures are 52.75, 53.4, 53.77, 54.5, 55, 55.5, 56, 56.5, 57, 57.5, and 58 K from bottom to top. **(b).** magnetoresistance calculated as $(R_{max}-R_{min})/R_{min}$ vs. temperature. R_{max} and R_{min} are MR values at the peak and at the dip near 0T respectively, see figure 1(c) Inset: Resistive transition in zero magnetic field (dotted line) for the same sample. **(c).** Solid circles: low field zoom of the R (H) loop at 55 K of the same sample as in Figure 1 (c). Solid line: hysteresis loop at 58 K (just at the superconducting onset). Stars are magnetization values obtained from polarized neutron reflectometry.

Large magnetoresistance (MR) peaks are observed whose relative height decreases when temperature is increased. Figure 1(b) shows that magnetoresistance, MR =(R_{max}-R_{min})/R_{min} , decreases with increasing temperature along the resistive transition (see inset of figure 1(b)) and it is abruptly suppressed at the resistive onset of the superconducting transition. This evidences that superconductivity plays a key role in the occurrence of this MR phenomenon and rules out a conventional GMR effect as observed in magnetic superlattices. Figure 1 (c) shows an enlarged view of the 55 K R(H) loop together with the hysteresis loop measured at 58 K, just at the superconducting onset. The hysteresis loop did not change appreciably when the measuring temperature was at the zero resistance value (50 K). The step- like hysteresis loops points to some degree of antiferromagnetic (AF) alignment between the two F layers, resulting from different switching fields of bottom and top layer layers as confirmed by polarized neutron reflectometry (PNR) [20](not shown). Stars in figure 1 (c) is the magnetization as calculated from PNR measurements. Interestingly, the MR peaks occur exactly in the field-interval where PNR detects antiferromagnetic alignment.

In addition, the size of the MR peaks was independent on whether the in plane current was parallel or perpendicular to the magnetic field. Figure 2 shows magnetoresistance peaks of the same sample of figure 1 measured at 55.5 K, with current in the plane of the layers, and directed parallel (line) and perpendicular (open symbols) to magnetic field. It can be observed that the size and shape of the peaks does not depend on the angle between magnetic field and current, ruling out explanations related to the anisotropic magnetoresistance (AMR) of the single ferromagnetic layers which, in fact, shows up when the temperature is raised above the superconducting onset.

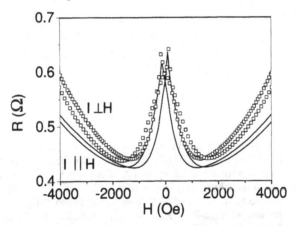

Figure 2. Resistance as a function of magnetic field, R(H) loops, at 55.5 K, of a F/S/F trilayer [LCMO (40 u.c.) / YBCO (15 u.c.) / LCMO (40 u.c.)] (same sample as in figure 1) with magnetic field parallel (solid line) and perpendicular (open symbols) to current.

It turns out that the magnetoresistance in the LCMO/YBCO/LCMO trilayers has many of the ingredients of the GMR in magnetic superlattices: it has actually giant values, it is independent of the current direction, and it depends on the relative orientation of the F layers. However, the magnetoresistance is not due to ordinary GMR since it is absent in the normal state of the YBCO layer and only occurs with the onset of superconductivity. Given that the resistance is increased with antiferromagnetic alignment of the F layers, this suggests that for the antiferromagnetically aligned magnetic layers the zero resistance critical temperature of the YBCO is reduced. Notice, that this effect is opposite to the result observed by Gu et al. in F/S/F trilayers based on conventional low-Tc superconductors and transition metal ferromagnets, where proximity effect yields higher Tc values when magnetic layers are AF aligned [17].

Increasing the thickness of the YBCO spacer results in a rapid decrease of the magnetoresistance and the effect is drastically reduced for thickness larger than 30 unit cells. Figure 3 shows results for a sample with 40 unit cells thick YBCO. Figure 3(a) shows magnetoresistance sweeps in which magnetoresistance peaks are visible only at the lowest temperatures. Figure 3(b) quantifies the magnetoresistance from minima and maxima as in figure 1(b). Figure 3(c) shows an enlarged view of the sweep at 67.5 K together with the hysteresis loop. Note that the hysteresis loop does not show now the step characteristic of the AF alignment between the magnetization of both ferromagnetic layers.

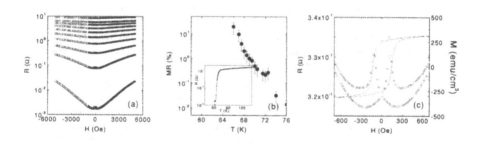

Figure 3. (a). Resistance as a function of magnetic field, R(H) loops, of a F/S/F trilayer [LCMO (40 u.c.) / YBCO (40 u.c.) / LCMO (40 u.c.)] at different temperatures along the resistive transition. Magnetic field was swept between –0.6 T and 0.6 T fields. Temperatures are 66, 67, 67.5, 68, 68.5, 69, 69.5, 70, 70.5, and 71 K from bottom to top. **(b).** magnetoresistance calculated as $(R_{max}-R_{min})/R_{min}$ vs. temperature. Inset: Resistive transition in zero magnetic field (dotted line) for the same sample. **(c).** Solid circles: low field zoom of the R (H) loop at 67.5 K of the same sample as in Figure 1 (c). Solid line: hysteresis loop at the superconducting onset.

We now discuss the origin of this MR in connection with the spin imbalance theory of Takahashi, Imamura, and Maekawa [18] although in the CIP geometry the spin accumulation concept does not apply. In the CIP geometry, equipotential planes are perpendicular to the layers, and current flows essentially parallel to the layers. However, since along the resistive transition (where our MR effect is observed) the resistance is finite, part of the current will be carried through the magnetic layers, where carriers will be frequently scattered into the superconducting layer. These electrons will cross the superconductor and will be strongly scattered at the interface with the other ferromagnetic layer if the alignment is AF. AF alignment therefore results in an increased number of quasiparticles in the superconductor what selfconsistently reduces the critical temperature. This yields the increased magnetoresistance at a given temperature for the magnetic field interval where AF alignment occurs. In the F configuration the strong interface scattering does not occur since incoming spins have the same orientation as the layer they approach to and the critical temperature remains unaltered. The decrease of the effect when the thickness of the superconductor is increased can be discussed in two different frameworks. On the one hand (evanescent) quasiparticles with energy smaller than the gap will diffuse with a characteristic length scale close to the Ginzburg Landau coherence length [25]. Blonder-Tinkham- Klapwijk (BTK) theory [26] predicts evanescent quasiparticle states over a distance 1.22 $\xi_{GL}(T)$, which diverges at Tc. On the other hand quasiparticles with energies larger than the gap (which are expected to be a larger fraction close to Tc) will diffuse into the superconductor keeping spin orientation over the spin diffusion length. Increasing the thickness of the superconductor strengthens the gap in the middle of the layer what reduces the number of quasiparticles and thus magnetoresistance. In the spin diffusion scenario, increasing the thickness of the superconductor will also (exponentially) reduce magnetoresistance, and magnetoresistance should vanish when the thickness approaches the spin diffusion length. The CIP geometry does not allow to establish which is the dominating mechanism although the small coherence length of the cuprates together with the d wave superconducting gap allowing low energy quasiparticle diffusion along the nodes at zero energy cost seem to favor the spin diffusion scenario.

In summary we have found large MR (up to 1600%) in ferromagnetic / superconducting heterostructures made of highly spin polarized LCMO and high-Tc superconducting YBCO. This novel MR is reminiscent of the GMR in metallic superlattices insofar as it depends on the relative orientation of the magnetic layers and is independent on the relative direction of current and field, but with much larger values. However, in contrast to traditional GMR, the magnetoresistance vanishes in the normal state of the YBCO and only occurs in the superconducting state. Furthermore, the magnetoresistance is opposite in sign to magnetoresistance effects observed in heterostructures based on low-Tc superconductors and transition metal ferromagnets. The possible origin of this MR is the depressed order parameter in the superconductor due to strong interface scattering at the F/S interface in the AF configuration.

ACKNOWLEDGMENTS

Work supported by MCYT MAT 2005-06024, Fundación Ramón Areces, CAM GR-MAT-0771/2004.

REFERENCES

On leave from Universidad del Quindio. Armenia. Colombia

[1] Z. Radovic, L. Dobrosavljevic-Grujic, A. I. Buzdin and J. R. Clem, Phys. Rev. B **38**, 2388 (1988). Z. Radovic, M. Ledvij, L. Dobrosavljevic-Grujic, A. I. Buzdin and J. R. Clem, Phys. Rev. B **44**, 759 (1991)

[2] I. Baladié and A. Buzdin, Phys. Rev. B **67**, 014523 (2003)

[3] V.V. Ryazanov, V.A. Oboznoz, A.Yu. Rusanov, A.V. Veretennikov, A.A. Golubov and J. Aarts, Phys. Rev. Lett. **86**, 2427 (2001).

[4] T. Kontos, M. Aprili, J. Lesueur, F. Genêt, B. Stephanidis, and R. Boursier Phys. Rev. Lett. **89**, 137007 (2002)

[5] C. Uher, R. Clarke, G.G. Zheng, and I.K. Schuller, Phys. Rev. B**30**, 453 (1984).

[6] Th. Mühge, N.N. Garif'yanov, Yu. V. Goryunov , G.G. Khaliullin, L.R. Tagirov, K. Westerholt, I.A. Garifullin, and H. Zabel, Phys. Rev. Lett. **77**, 1857 (1996).

[7] J. Aarts, J. M. E. Geers, E. Brück, A. A. Golubov, and R. Coehorn, Phys. Rev. B **56**, 2779 (1997)

[8] S. Kaneko, U. Hiller, J.M. Slaughter, Charles M. Falco, C. Coccorese, and L. Maritato, Phys. Rev B**58**, 8229 (1998)

[9] G. Verbanck, C.D. Potter, V. Metlusko, R. Schad, V.V. Moshchalkov, and Y. Bruynseraede, Phys. Rev. B**57**, 6029 (1998)

[10] L. Lazar, K. Westerholt, H. Zabel, L. R. Tagirov, N. N. Garifyanov, V. Goryunov, G. G. Khaliullin, Yu. I, A. Garifullin, Phys. Rev. B**61**, 3711 (2000)

[11] G. Jakob, V. V. Moshchalkov, and Y. Buynseraede, Appl. Phys. Lett. **66**, 2564 (1995)

[12] P. Przyslupski, S. Kolesnik, E. Dynovska, T. Skoskiewicz and M. Sawicki, IEEE Trans. Appl. Superconductivity **7**, 2192 (1997)

[13] C. A. R. Sá de Melo, Phys. Rev. Lett **79**, 1933 (1997); ibid. Phys. Rev. B**62**, 12303 (2000)

[14] P. Prieto, P. Vivas, G. Campillo, E. Baca, L. F. Castro, M. Varela, C. Ballesteros, J. E. Villegas, D. Arias, C. Leon and J. Santamaria, J. Appl. Phys. **89**, 8026 (2001)

[15] H.-U. Habermeier, G. Cristiani, R. K. Kremer, O.I. Lebedev and G. Van Tendeloo, Physica C **354**, 298 (2001)

[16] Z. Sefrioui M. Varela, V. Peña, D. Arias, C. Leon, J. Santamaria, J. E. Villegas, J. L. Martinez, W. Saldarriaga and P. Prieto Appl. Phys. Lett. **81**, 4568 (2002)

[17] J. Y. Gu, C.-Y. You, J. S. Jiang, J. Pearson, Ya. B. Bazaliy, and S. D. Bader, Phys. Rev. Lett. **89**, 267001 (2002)

[18] S. Takahashi, H. Imamura, and S. Maekawa, Phys. Rev. Lett. **82**, 3911 (1999).

[19] C. D. Chen, Watson Kuo, D. S. Chung, J. H. Shyu, and C. S. Wu, Phys. Rev. Lett. **88**, 47004 (2002)

[20] V.Peña Z. Sefrioui, D. Arias, C. Leon, J. Santamaria, J. L. Martinez, S. G. E. Te Velthuis and A. Hoffmann, Phys. Rev. Lett. **94**, 057002 (2005)

[21] Z. Sefrioui, D Arias, J. E. Villegas, M. Varela, V. Peña, P. Prieto, C. León, J. L. Martínez y J. Santamaría. Phys. Rev. B **67**, 214511 (2003)

[22] V. Peña, Z. Sefrioui, D Arias, C. León, J. L. Martínez and J. Santamaría. Eur. Phys. Jour. B **40**, 479 (2004)

[23] V. Peña, Z. Sefrioui, D. Arias, C. Leon, J. Santamaria, M. Varela, S. J. Pennycook, and J. L. Martinez. Phys. Rev. B **69**, 224502 (2004)

[24] I. K. Schuller, Phys. Rev. Lett. **44**, 1597 (1980); W Sevenhans, M. Gijs, Y. Bruynseraede, H, Homma and I. K. Schuller, Phys. Rev. B **34**, 5955 (1986); E. E Fullerton, I. K. Schuller, H. Vanderstraeten and Y. Bruynseraede, ibid.. B **45**, 9292 (1992); D. M. Kelly, E. E Fullerton, J. Santamaria and I. K. Schuller, Scripta Met. Mat. **33**, 1603 (1995)

[25] J. J. Gu, J. A. Caballero, R. D. Slater, R. Loloee, and W. P. Pratt, Jr., Phys. Rev. B **66**, 140507R (2002)

[26] G. E. Blonder, M. Tinkham, and T. M. Klapwijk Phys. Rev. B **25**, 4515 (1982)

Mater. Res. Soc. Symp. Proc. Vol. 887 © 2006 Materials Research Society   0887-Q11-03

Double Layer Processes of LBMO/YBCO and Crystalline Degradations

Hong Zhu[1], Masanori Okada[1], Hidetaka Nakashima[1], Ajay K. Sarkar[1], Hirofumi Yamasaki[3], Kazuhiro Endo[2], Tamio Endo[1,2]

[1]Faculty of Engineering, Mie University, Tsu, Mie 514-8507, Japan
[2]NeRI, National Institute of Advanced Industrial Science and Technology, Tsukuba, Ibaraki 305-8568, Japan
[3]ETRI, National Institute of Advanced Industrial Science and Technology, Tsukuba, Ibaraki 305-8568, Japan
Corresponding author: Tamio Endo, endo@elec.mie-u.ac.jp

ABSTRACT

Double-layers of YBCO on LBMO and LBMO on YBCO were fabricated by ion beam sputtering technique. In the case of YBCO/LBMO, excellent a- and c-oriented overlying YBCO can be grown on underlying LBMO at 600- 650 °C. Crystallinity of the overlying YBCO is nearly the same with that of YBCO single-layers on MgO and LaAlO$_3$ (LAO) substrates. Mosaicity of YBCO overlayer is much better than that of YBCO single-layers on MgO. It is noticed that the crystallinity of underlying LBMO can be improved, and the mosaicity is not degraded during the double-layer deposition. However, the inferiority is that surface roughness of the double-layer is much degraded. In the case of LBMO/YBCO, the excellent crystalline LBMO can be grown on the underlying a- and c-oriented YBCO at 650-700 °C. LBMO/a-YBCO clearly shows XRD peak separation but LBMO/c-YBCO does not show the peak separation.

Time dependences of the crystalline degradations on YBCO/LBMO double-layers and corresponding single-layers are presented as well. The crystallinity of YBCO/MgO is easily degraded with time, but that of YBCO/LAO does not show the degradation. LBMO is stable if it is covered with YBCO overlayer.

INTRODUCTION

For nearly 20 years, high temperature superconductivity (HTS) in copper oxides and colossal magnetoresistance (CMR) in manganese oxides have drawn considerable attention to transition metal oxides with perovskite structures. Recently, CMR/HTS double-layer film, as one of the typical samples in so-called ferromagenet (Ferro)-supreconductor hybrids [1,2], has become an interesting topic in fundamental researches, such as vortex pinning and spin-polarized carrier injection, etc. In field of practical application, Ferro/HTC double-layer films have also been proposed as one of promising candidates for next generation of tunable microwave filter after realization of HTS microwave filter by the late 1990s [3-5].

In order to obtain the excellent double-layer films, knowledges on the correlation between the layers, such as underlayer quality dependence of overlayer and degradations of underlayer during overlayer deposition, are essentially important. In this paper, we report epitaxial growth of $YBa_2Cu_3O_x$ (YBCO)/La(Ba)MnO$_3$ (LBMO) and LBMO/YBCO double-layer films. The epitaxial qualities of double-layers are compared with that of the corresponding single-layers. The crystalline degradations of the underlayers after the overlayer deposition are also investigated.

EXPERIMENTAL DETAILS

The films were prepared by means of ion beam sputtering (IBS) technique [6], where composite particles sputtered from a polycrystalline target by 4 keV Ar$^+$ ions were deposited on heated substrates with temperature (T_S) varied from 600 to 700°C. Molecular oxygen (ML) with various partial pressures (P_O=0.1~3 mTorr) was supplied during the deposition. To compare effects of active oxygen, some films were supplied by plasma oxygen (PL) produced in a plasma source with the AC discharge voltage of 1 kV and current of 10 mA at 60 Hz. Two series of double-layer films were prepared, i.e., YBCO on LBMO (YBCO/LBMO) and LBMO/YBCO on the opposite. For the former one, the LBMO underlayer was firstly epitaxially grown on LaAlO$_3$ (LAO) (001) substrate, and the second step was the deposition of YBCO overlayer on it. For comparison, YBCO films were also deposited directly on LAO and MgO substrates simultaneously during the double-layer deposition [7]. For the second serious of LBMO/YBCO, the process was similar to that of YBCO/LBMO double-layer and a main difference was substitution of MgO substrate for LAO.

To characterize crystallinities of the double- and single-layer films, X-ray diffraction (XRD) was carried out using a Philips diffractometer with Cu K_α radiation. Plane-distance crystallinity (simply called "crystallinity" hereafter) and plane-direction crystallinity (mosaicity) were evaluated by full width at half maximum of θ-2θ peak (Δ_θ), and ω-scan rocking curve (Δ_ω), respectively. Surface morphologies of the films were characterized by atomic force microscopy (AFM), and root mean square (rms) surface roughness R_{sf} was obtained from the AFM image.

RESULTS AND DISCUSSION

YBCO/LBMO double-layer films

Figure 1 shows a set of the typical XRD patterns of films. From top to bottom panel, they are YBCO single-layers on MaO and LAO substrate, YBCO/LBMO/LAO double-layer, and LBMO underlayer just before the double-layer deposition, respectively. It can be seen that LBMO underlayer, deposited at T_S=700°C and P_O=1 mTorr with PL supply, is a highly epitaxial film. YBCO single-layers, deposited simultaneously with the double-layer at T_S=650°C and P_O=1.5 mTorr with PL supply, are epitaxial films with a-orientation. As for YBCO/LBMO double-layer film, XRD pattern shows a detectable YBCO (200) peak besides strong LBMO (00n) peaks. It is noticeable that an intensity of YBCO (200) peak is rather weak, however, it becomes higher when the sample is titled a little in XRD measurement (not shown here). The result suggests that the overlaying a-YBCO grows slightly titled from the substrate plane due to the surface fluctuation of LBMO underlayer.

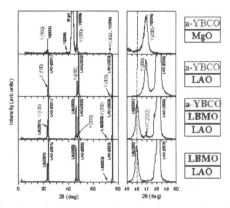

Figure 1. XRD patterns for YBCO single-layers on MgO and LAO substrates, YBCO/LBMO/LAO double-layer, and LBMO underlayer just before the double-layer deposition, from top to bottom.

To investigate influences of the double-layer depositions on the crystallinity, we plotted Δ_θ values of YBCO (200) and LBMO (002) peaks in figure 2 for various layers (1-4) with four different deposition conditions. It is clear that Δ_θ for YBCO single-layers on LAO (2) are smaller than those on MgO substrates (1), simply due to better lattice matching on LAO. As for the one sample of overlying YBCO in the double-layer films (3), a value of Δ_θ denoted by a symbol-Y, is slightly larger than those for YBCO single-layers on LAO, but is almost the same with those on MgO. It indicates that YBCO overlayer can be grown on LBMO underlayer with high crystallinity as on MgO and LAO substrates. It is also known that Δ_θ for LBMO underlayers (3)

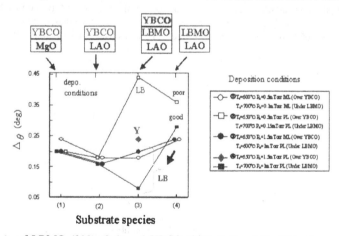

Figure 2. Δ_θ of LBMO (002) (3-4) and YBCO (200) (1-2 and 3) diffraction peaks for the single-layers and double-layers under various deposition conditions.

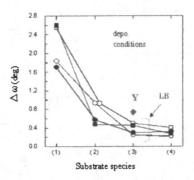

Figure 3. Δ_ω of LBMO (002) and YBCO (200) rocking curves for the single-layers YBCO and YBCO/LBMO double-layers. Refer to figure 2 for substrate species and depositions.

are decreased typically from those before the double-layer deposition (4). This suggests that the crystallinity of the underlaying LBMO is improved after YBCO overlayer deposition.

Concerning the mosaicity, we show in figure 3 Δ_ω of YBCO (200) or LBMO (002) rocking curves for YBCO single-layers (1-2), LBMO single-layers before the double-layer deposition (4), and the double-layer films (3). The mosaicity of YBCO single-layers on LAO is again much better than that on MgO substrate as expected. In the double-layer film, it is incredible that the mosaicity of YBCO overlayer indicated by a Symbol-Y is almost the same with that of the films having highly lattice-matching with LAO substrate. As for the underlying LBMO (3), the mosaicity is not affected after YBCO overlayer deposition.

Figure 4 shows the surface roughness (R_{sf}) of the single-layer films and double-layer films deposited under various conditions. For YBCO single-layers, the minimum surface roughnesses can be obtained under the condition ③, which are 0.33 nm with a-orientation on LAO and 1.04 nm with c-orientation on MgO substrate, respectively. The values approximate to the a- and c-lattice parameters of YBCO, indicating ultimately smooth surface with 1-unit-cell fluctuation.

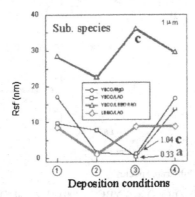

Figure 4. Surface roughnesses of the single- and double-layers for various deposition conditions.

However, the surface roughnesses of the double-layer films are much larger than those of YBCO single-layers and LBMO underlayers. In general, the variation of the double-layer film roughness follows the parallel tendency as that of LBMO underlayer, suggesting the underlayer roughness plays an important role in the double-layer roughness. Then improvement of LBMO underlayer smoothness is essentially necessary to obtain the well-smooth double-layer film.

LBMO/YBCO double-layer films

To compare the crystallinity and surface evolutions in two configurations, the opposite deposition process was also employed, which is YBCO underlayer deposition on MgO substrate followed by LBCO overlayer deposition. The detailed deposition conditions for various LBMO/YBCO samples are given in table 1. During the overlayer deposition, LBMO single-layers were deposited on MgO and LAO substrates simultaneously. Figure 5 shows a group of

Table 1. Deposition conditions for LBMO overlayer (upper) and YBCO underlayer (lower).

①	②	③	④	⑤
$T_S=700°C$, $P_O=2$ mTorr, ML	$T_S=600°C$, $P_O=1$ mTorr, PL	$T_S=700°C$, $P_O=2$ mTorr, ML	$T_S=650°C$, $P_O=1$ mTorr, PL	$T_S=650°C$, $P_O=1$ mTorr, PL
$T_S=600°C$, $P_O=3$ mTorr, ML	$T_S=600°C$, $P_O=1.5$ mTorr, ML	$T_S=600°C$, $P_O=2.5$ mTorr, ML	$T_S=700°C$, $P_O=1$ mTorr, PL	$T_S=600°C$, $P_O=0.2$ mTorr, ML

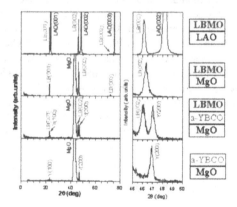

Figure 5. XRD patterns for LBMO singl-layers on LAO and MgO substrates, LBMO/YBCO double-layer, and YBCO underlayer before double-layer deposition. Condition ③.

XRD patterns for LBMO single-layers and LBMO/YBCO double-layer under the deposition condition ③. For comparison, XRD pattern of YBCO underlayer collected just before the double-layer deposition is also shown in the bottom panel of figure 5. It is known that YBCO underlayer before the double-layer deposition is the a-orientated epitaxial film. With regard to the double-layer film, LBMO overlayer can be epitaxially grown on the a-YBCO underlayer as well. LBMO (002) peak is clearly separated from YBCO (200) peak because of the distinct difference in the lattice parameters (~0.390 nm and ~0.385 nm, respectively). Compared with the result in YBCO/LBMO structure as shown in figure 1, the (200) peak intensity of underlying YBCO is considerably larger, indicating YBCO underlayer in LBMO/YBCO structure is not titled even after LBMO overlayer deposition.

Figure 6 shows the LBMO (200) and YBCO (200) XRD peak positions ($2\theta_p$) for the single-

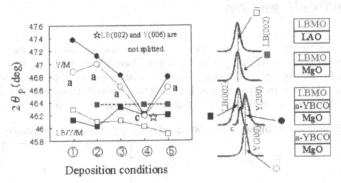

Figure 6. Peak positions of LBMO (002) and YBCO (200) reflections for the single- and double-layer films deposited under various conditions.

and double-layer films deposited under various conditions. After the deposition of LBMO overlayer, YBCO (200) peaks generally shift to higher angles except for condition ④, corresponding to the shrinkage of a-axis in YBCO. It is most likely due to the oxygen evolution in the underlying YBCO during the overlying LBMO deposition at higher T_S. In condition ④, YBCO underlayer with c-orientation was deposited at T_S=700°C, whereas LBMO overlayer was deposited at T_S=650°C. The double-layer does not show the peak separation. The c-phase peak of (006) is located at lower 2θ angle than the a-phase peak of (200) due to longer c/3 than a-b axis, then (006) peak is close to LBMO (002) peak, showing no peak separation.

For LBMO single-layer and overlayer, the positions of (002) peak are not so sensitive to the deposition condition, indicating that the lattice match is more important to the lattice parameter of LBMO film.

Aging effects of YBCO/LBMO double-layer films

A stability of high temperature superconductors, especially in their thin film forms, is an important factor for practical applications. So we investigated aging effects of YBCO/LBMO double-layers and the corresponding single-layer films. The crystallinity and surface of the films were reevaluated 25 months later after deposition, and they were compared with the results of the as-grown films.

Figure 7 shows the ratio of Δ_θ for the aged samples over the as-grown samples for various

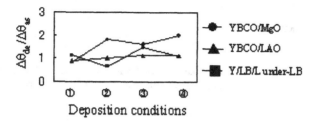

Figure 7. The ratios of Δ_θ for aged samples over the as-grown samples; YBCO single-layers on MgO and LAO substrates, and underlying LBMO in YBCO/LBMO double-layer films.

deposition conditions. It is clear from figure 7 that Δ_θ of YBCO single-layers on MgO substrate increase almost by 2 times after 25 months, indicating a serious degradation in the crystallinity. In the case of LAO substrate, however, the ratios of Δ_θ remain unity. Then the crystallinity of YBCO single-layers on LAO is not degraded. The results clearly suggest that YBCO films on more lattice-matched substrate (LAO) have much longer life as compared to those on less lattice-matched substrate (MgO). YBCO on MgO has much more grain boundaries than that on LAO, then oxygen effusion is more serious, leading to larger degradation. Another reason is that YBCO on MgO is attacked by water easier through grain boundaries, leading to easier dissolving. As for the underlying LBMO in the double-layer films, the ratios of Δ_θ are around unity, indicating that the crystallinity is not degraded after such a long time. It is thought that LBMO does not have such serious oxygen effusion and dissolving. Another simple extrinsic reason is that the underlying LBMO is protected by the overlying YBCO layer.

CONCLUSIONS

The epitaxial LBMO and YBCO double-layer films were successfully fabricated on MgO and LAO substrates using the ion beam sputtering technique. In YBCO/LBMO structure, the crystallinity of the overlying YBCO is not poor but rather good. Moreover, its mosaicity is excellent and much better than that of the single-layer YBCO on MgO. As for LBMO underlayer, the crystallinity is improved and the mosaicity remains unchanged after the deposition of YBCO overlayer. The inferiority for such structure is the degradation of the surface smoothness after the double-layer deposition.

In LBMO/a-YBCO structure, the both of underlayer and overlayer show fine XRD peaks. Then the good crystalline overlying LBMO can be grown on a-YBCO. It shows well separated XRD peaks. However, LBMO/c-YBCO double-layer does not show the peak separation because the c-phase peak is located at lower 2θ angle than the a-phase, resulting in the overlapping of LBMO and c-YBCO peaks.

The crystallinity of YBCO single-layer on MgO is considerably degraded with time, however, that on LAO is not degraded. The reason for this difference must be grain boundaries. YBCO on the mismatched MgO should have more grain boundaries. For the underlying LBMO, the crystallinity remains unchanged after a long time. Then LBMO/YBCO structure may be promising for the double-layer devices.

ACKNOWLEDGMENTS

One of the authors, H. Zhu, would like to thank Japan Society for the Promotion of Science (JSPS) for supporting his researches in Japan. This work is supported by Gran-in-Aid from JSPS.

REFERENCES

1. I. F. Lyuksyutov, and V. L. Pokrovsky, *Advances in Physics* **54**, 67 (2005).
2. J. Albrecht, S. Soltan, and H.-U. Habermeier, *Phys. Rev. B* **72**, 092502 (2005).
3. A. T. Findikoglu, Q. X. Jia, X. D. Wu, G. J. Chen, T. Venkatesan, and D. W. Reagor, *Appl. Phy. Lett.* **68**, 1651 (1996).
4. A. Lauder, K. E. Myers, and D. W. Face, *Adv. Mater.* **10**, 1249 (1998).
5. Z. Y. Shen, C. Wilker, P. Pang, D. W. Face, C. F. Carter III, and C. M. Harrington, *IEEE Trans. Appl. Supercond.* **7**, 2446 (1997).
6. M. Tada, J. Yamada, V. V. Srinivasu, V. Sreedevi, H. Kohmoto, A. Hashizume, Y. Inamori, T. Tanaka, A. harrou, J. Nogues, J. S. Munoz, J. M. Colino, and T. Endo, *J. Crystal Growth* **229**, 415 (2001).
7. T. Endo, H. Kohmoto, S. Iwasaki, M. Matsuo, M. Matsui, Y. Kurosaki, H. Nakanishi, and K. Niwano, *New Materials*, (ARCI, Hyderabad, 2002) pp. 205-223.

Mater. Res. Soc. Symp. Proc. Vol. 887 © 2006 Materials Research Society   0887-Q11-04

Superconductors in confined geometries

Z. L. Xiao*, Y. S. Hor, U. Welp, Y. Ito*, U. Patel*, J. Hua*, J. Mitchell, W. K. Kwok, and G. W. Crabtree
Materials Science Division, Argonne National Laboratory, Argonne, IL 60439, U.S.A.
*Also at the Department of Physics, Northern Illinois University, DeKalb, IL 60115, U.S.A.

ABSTRACT

The synthesis of nanoscale superconductors with controlled geometries is extremely challenging. In this paper we present results on synthesis and characterization of one-dimensional (1D) $NbSe_2$ superconducting nanowires/nanoribbons. Our synthesis approach includes the synthesis of 1D $NbSe_3$ nanostructure precursors followed by nondestructive and controlled adjustment of the Se composition to formulate $NbSe_2$. The morphology, composition and crystallinity of the synthesized 1D $NbSe_2$ nanostructures were analyzed with scanning electron microscopy, x-ray diffraction and transmission electron microscopy. Transport measurements were carried out to explore the electronic properties of these confined superconducting nanostructures.

INTRODUCTION

One dimensional (1D) nanostructures in the form of free-standing wires, ribbons and tubes have been the subject of intensive research in recent years due to their novel properties and intriguing applications [1-4]. Most of the reported work in this area has focused on nanostructures of semiconductors [5,6], noble metals [4,7,8], dielectric [3,9,10] and magnetic materials [11,12]. Free-standing 1D superconducting nanostructures will be highly desirable in future electronic nanodevices as interconnects since they circumvent the damaging heat produced by energy dissipation in a normal nanoconductor whose high resistance is inversely proportional to its cross-section area. Furthermore, when Cooper pairs are squeezed into a small volume, their wave functions are strongly modified, and therefore, nanoscale superconductors are expected to exhibit properties that are different from bulk materials. Due to the confinement of Cooper pairs, new phenomena such as vortex-antivortex pairs and fractional flux quanta have been observed in 2D superconducting microstructures [13-16]. With further reduction of the dimensionality from 2D to 1D, a fundamental question arises as to how the collective properties of superconductors are affected. This issue is also of practical importance in defining the size limit of a superconducting wire with regards to potential applications in electronic circuits. So far, however, there are only a limited number of studies dealing with the synthesis and characterization of free-standing 1D superconducting nanostructures. By utilizing carbon nanotubes (individual or bundle) as templates, 1D superconducting MoGe [17] and Nb [18,19] nanowires as small as a few nanometers have been fabricated. The majority of the publications report on electrodeposition synthesis of lead (Pb) and tin (Sn) nanowires by using nanochannels in porous membranes such as track-etched polycarbonate [20] and anodic aluminum oxide [21] as templates. Progress has also been made in synthesizing Pb nanowires by thermal decomposition of lead acetate in ethylene glycol [22] and directly growing Pb nanowires onto graphite substrates at high electrodeposition reduction potentials [23]. Superconducting nanowires of MgB_2 have also been fabricated by reacting B nanowires with vapors of Mg

[24,25]. Here we report synthesis and characterization of a new class of 1D nanostructures which include nanowires and nanoribbons of niobium diselenide ($NbSe_2$) superconductors. $NbSe_2$, in bulk crystal form has been used in recent years as a model system to study vortex physics due to its low pinning strength, unique atomic flat surface and intriguing 'peak effect' which represents itself as a peak in the temperature/magnetic field dependence of the critical current just before it vanishes [26,27]. This provides a useful reference for the properties of 1D confined nanostructures. For example, we observed an enhancement of the critical field and the disappearance of the peak effect in 1D geometry. Experimental details on synthesis, structural characterization and superconducting properties will be presented below.

EXPERIMENTAL DETAILS

The essential part of our synthesis procedure is to transform $NbSe_3$ nanostructures into superconducting $NbSe_2$, while preserving their shapes. For the synthesis of $NbSe_3$ nanostructure precursors, stoichiometric quantities of high purity (>99.99%) niobium and selenium powders with particle sizes of several micrometers were combined. The mixture was ground and sealed in an evacuated quartz ampoule of 17 mm in inner diameter and ~15 cm in length, after purging repeatedly with high purity Ar gas to ensure an oxygen free environment. By sintering the mixtures at appropriate temperatures $NbSe_3$ nanostructures were synthesized. In the conversion process, we placed $NbSe_3$ nanostructures on one end of the quartz tube and a stoichiometric amount of Nb powder (see below for details) at the center of the tube, approximately 5 cm away from the $NbSe_3$ nanostructures. The same procedure as that for $NbSe_3$ was applied in evacuation and sealing of the quartz tubes. Ampoules were placed in a Lindberg furnace and heated up to various temperatures at a rate of 3 °C/min and maintained at the synthesis temperatures for various period of sintering before cooling down to room temperature at 2 °C/min. We calibrated the furnace temperature where the quartz ampoule is located with a type-K (Chromel-Alumel) thermocouple. The temperature stability is about ± 2 °C. Powder x-ray diffraction (XRD) analysis with a monochromatic CuK_α ($\lambda=1.540598$ Å) radiation source was used to determine the phase of the product. The morphological analyses were performed with a scanning electron microscope (SEM) (Hitachi S- with selected 4700-II). A transmission electron microscope (TEM) (FEI Tecnai F20ST) equipped with selected area electron diffraction (SAED) attachments was used to obtain structural information on individual nanowires. A SQUID magnetometer and transport measurements were used to characterize the superconductivity of the converted bundle of $NbSe_2$ nanostructures (in the form of nanowire bundle) and individual nanowire, respectively.

RESULTS AND DISCUSSION

Niobium triselenide ($NbSe_3$) crystals possess a pseudo one-dimensional structure with infinite chains of triangular prismatic units parallel to the b-axis [28,29]. The chains with strong bonding are separated by relatively large distances. As a result, $NbSe_3$ single crystals usually grow in the form of whiskers/ribbons with length up to millimeters and smallest cross-section dimension of a few micrometers to hundreds of micrometers [30]. Typical processes for fabricating nanoscale $NbSe_3$ samples include two-steps. First, bulk $NbSe_3$ ribbons are synthesized through a standard crystal growth procedure [30] and nanoscale structures are derived from bulk ribbon-like structures with either e-beam lithography followed by plasma etching [31] or ultrasonic shaking that exploits the weak bonds between chains [32]. These approaches are either complicated [31]

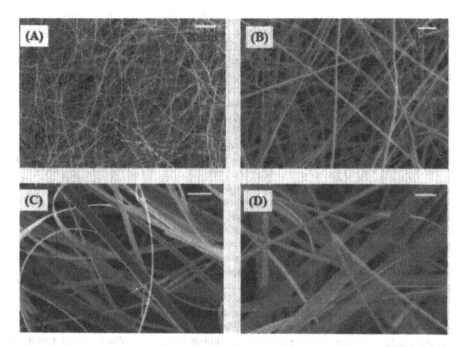

Figure1. Scanning electron microscopy (SEM) micrographs of NbSe₃ nanostructures synthesized by direct reaction of Nb and Se powders in an evacuated quartz tube at 700 °C. (a) low magnification image shows the length of the mesowires can be as long as several millimeters, (b) image from the same sample but with higher magnification to identify the diameter and morphology of the mesostructures which seem to be cylindrical and with diameters between a few tens of nanometers to a few micrometers, (c) and (d) show appearance of mesoribbons.

or can induce defects [32]. Furthermore, these methods can only produce a limited quantity of nanoscale samples. Through direct reaction of stoichiometric Nb and Se powders in an evacuated quartz tube [33], we obtained one-dimensional NbSe₃ nanostructures with a one-step procedure. As demonstrated by the SEM micrographs shown in Fig.1, the synthesized NbSe₃ nanostructures are tens of micrometers to a few millimeters in length. Their typical cross-section is rectangular. Following conventional terminology [34], nanowires (nanoribbons) will be used to refer to the nanostructures with width smaller (larger) than 5 times the thickness. The widths of the synthesized nanostructures range from ~20 nm to ~700 nm, although occasionally, bundles with width up to micrometers can also be observed. The nanoribbons can be as thin as a few nanometers.

We found that sintering temperature plays a crucial role in NbSe₃ nanostructure growth. To study temperature effects, mixtures of Nb and Se powders were heated in an evacuated quartz tube to various temperatures between 580 °C and 950 °C. We used identical quartz ampoules and equivalent amounts of starting material for each run. In all cases, only the sintering temperature was varied while heating and cooling profiles were unchanged. At temperatures above 600 °C, we observed that the ampoule filled with a brownish Se vapor. The higher the temperature, a thicker Se vapor was observed. SEM imaging shows that nanowires and nanoribbons begin to form when the temperature reaches 610 °C but remain lower than 700 °C. At higher temperatures, particles or rod-like objects with size of micrometers appear. X-ray diffraction (XRD) analyses indicate that the nanowires and nanoribbons are single-phase $NbSe_3$ with lattice constants consistent with the monoclinic structure reported in the literature while products obtained at temperatures above 700 °C are a mixture of $NbSe_2$ and Nb_2Se_9 (PDF 29-0950) [33].

We conducted transmission electron microscopy (TEM) and selected area electron diffraction (SAED) characterization on a nanoribbon with a width of ~ 230 nm and thickness on the order of 20 – 30 nm. The SAED pattern agrees with literature values for single crystals [35]. SAED patterns from the entire length of the ribbon display the same patterns except for slight diffraction conditions due to local bending, indicating that the ribbon is a single crystal. The long axes of the $NbSe_3$ nanowires and nanoribbons are parallel to the b axis while the smallest dimension of the nanoribbons is along the a axis. More data on morphological, structural characterizations and electronic properties can be found in Ref.33.

There are two critical parameters in the conversion of $NbSe_3$ into $NbSe_2$ nanostructures: the annealing temperature and the annealing atmosphere. The temperature should be high enough to enable the conversion reaction to occur but should not be too high as to destroy the shape of the $NbSe_3$ nanostructures. Experimentally, we found that the selenium in $NbSe_3$ can be extracted when annealed in non-selenium atmosphere at temperatures above 300 °C. A simple way to form $NbSe_2$ mesostructures will then be to anneal $NbSe_3$ in an inert gas such as argon. However, it is extremely difficult to determine the annealing conditions necessary to obtain the exact selenium composition of Se_2, because the final runaway product of this annealing approach will lead to Nb (or niobium oxide due to the extremely high reactivity of niobium with the residual oxygen in the annealing gas) when all the selenium is leached from the starting $NbSe_3$ structure.

We discovered a more controlled way to synthesize $NbSe_2$ nanowires and nanoribbons by controlling the annealing atmosphere, i.e., by sintering $NbSe_3$ nanostructures with an appropriate amount of Nb in a sealed quartz tube [36]. The reaction and the stoichiometry change can be expressed as $2NbSe_3 + Nb = 3NbSe_2$. The selenium bleached from $NbSe_3$ on one end of the quartz tube reacts with the Nb on the other end and the final product is expected to be only $NbSe_2$. Experiments have shown that the direct reaction of Nb and Se produces granular $NbSe_2$ powders. Hence, in order to preserve the morphology of the $NbSe_3$ nanostructure, we need to isolate it from the Nb powder in the quartz tube to prevent contamination of the converted $NbSe_2$ mesostructures with the $NbSe_2$ by-product powders. Since there are no other stable phases between $NbSe_3$ and $NbSe_2$, we expect the transformation of $NbSe_3$ to $NbSe_2$ to occur at the surface of the $NbSe_3$ nanostructure with the formation of a $NbSe_2$ shell. With increasing annealing time, the thickness of the shell increases inwards and displaces the $NbSe_3$ core. If the annealing time is long enough, the entire $NbSe_3$ nanostructure will be completely converted into $NbSe_2$ a nanowire or nanoribbon. This converted products synthesized at various temperatures and annealing times were analyzed with SEM, XRD, TEM, magnetization (SQUID) and transport measurements. SEM imaging determined that the shape of the $NbSe_3$ nanostructures remained unchanged if the annealing temperature is below 700 °C while particles and rod-like objects of micrometer size appeared at higher temperatures. XRD analysis, as presented in Fig.2 for the converted product at 700 °C, demonstrated that the annealed 1D nanostructures are pure

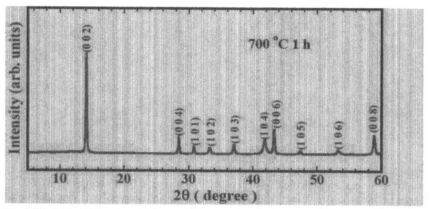

Figure 2. X-ray diffraction (XRD) pattern of NbSe₃ nanostructures annealed in an evacuated quartz tube sealed with an appropriate amount of Nb powder. The annealing temperature and time are 700 °C and 1 hour, respectively. The indexed peaks are for the NbSe₂ phase.

NbSe₂ when annealed at temperatures between 630 °C and 700 °C. At temperatures above 700 °C only NbSe₂ and Nb₂Se₉ phases can be identified in the annealed products. This indicates that in the temperature range of 630 ~ 700 °C, NbSe₃ nanostructures can be converted into NbSe₂ with no shape-change. High resolution TEM analysis and selected area electron diffraction also confirms the phase conversion from NbSe₃ to NbSe₂. However, high-density defects and grain boundaries were also detected [36], probably induced by the rearrangement of atoms during the crystalline structure change from a chain-like structure in NbSe₃ to the layered structure of NbSe₂.

The first and foremost characterization of a superconducting nanostructure is to investigate its superconducting transition. A quick examination of the superconductivity of the synthesized NbSe₂ nanostructures was performed with SQUID magnetization measurement. At a magnetic field of 10G, a transition from a normal to superconducting state was observed at about 7.20 K, the typical critical temperature of pure NbSe₂ bulk crystals. A more convincing experiment is to carry out transport measurements of individual nanostructures. We first conducted a quick two-probe transport measurement to determine the normal-superconducting phase boundary. The left panel of Fig.3 shows resistance versus temperature curves of a single NbSe₂ nanowire with a diameter of 800 nm at various applied magnetic fields perpendicular to the substrate. The abrupt drop in the resistance with decreasing temperature confirms the existence of superconductivity in the NbSe₂ nanowire. For this particular wire, a transition temperature of 7.05 K can be identified in the absence of an external magnetic field. The slight difference in the critical temperatures measured with magnetization and transport is probably due to the variation in the critical temperature within a bundle of nanostructures. Magnetization, being a global measurement, identifies the highest critical temperature in the bundle of nanostructures. From these resistance versus temperature curves, one can determine a magnetic-field versus temperature phase diagram of the normal and superconducting state. As shown in the right panel of Fig.3, this diagram gives

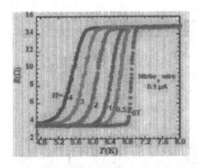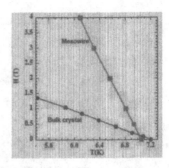

Figure 3. Left panel: resistance versus temperature curves of an 800 nm NbSe$_2$ nanowire obtained at various magnetic fields. Right panel: The magnetic field – temperature (H-T) diagram of this nanowire. The critical field data for bulk NbSe$_2$ crystal with magnetic field parallel to c-axis (provided by Prof. Eva Andrei at Rutgers University) are also shown for comparison.

a value of 4.2 T/K for dH_c/dT which is larger by a factor of six and two in comparison with the bulk values in the c-direction and a-b plane [37], respectively. It is predicted that the critical field H_c^* is larger than the bulk value of H_c for a superconductor with dimension of the order of the effective penetration depth λ_e. Such an enhancement of the critical field in small samples results from the reduced free energy density of the superconducting state due to a more significant effect of the field penetration in a small specimen. For a long cylindrical mesowire, the theoretically predicted relationship between H_c^* and H_c is $H_c^*/H_c = 8\lambda_e/d$, where d is the diameter of the mesowire. This has been confirmed in Pb nanowires [38]. For the measured NbSe$_2$ nanowire, a two-fold enhancement in the critical field compared to its bulk value in the a-b plane is consistent with the prediction of a factor of 2.3 for H_c^*/H_c, assuming that the field in our experiment is parallel to the a-b plane and the penetration depth is 230 nm [37].

As discussed above, NbSe$_2$ single crystals have been used intensively to study phase transition of vortex matter. Intriguing phenomena observed in this system include the peak effect and metastable states. Both are believed to be associated with the first-order transition of the vortex matter [26,39]. The investigation of the phase transition of vortex matter in nanoscale superconductors can be very useful in understanding the mechanism of the peak effect, e.g. the role of sample edges [39].

The four-probe transport measurement is a convenient way to detect the phase transition of vortex matter, for example, by measuring the current-voltage (I-V) characteristics at various temperatures and magnetic fields. By applying photolithography techniques we were able to make four contacts to a NbSe$_2$ nanowire with diameter of 700 nm and length of about 30 micrometers. Preliminary results, as demonstrated by the I - V curves in Fig.4, indicate that the phase transition of the vortex matter in NbSe$_2$ nanowires is remarkably different from that reported in bulk crystals. For example, there is no peak in the critical current (defined with a voltage criterion of 1 μVolt) versus magnetic field ($I_c \sim H$) curve i.e. the peak effect does not exist in this nanowire. The curvature change (in log-log plot) also resembles what is observed in high-T_c superconductors [40] and interpreted as a vortex glass transition [41,42]. However, a vortex glass state is supposed to exist only in superconductors with high-density point defects, e.g. high-T_c superconductors in which high-density point defects can be induced by oxygen vacancies. On the other hand, such a curvature change can also be induced by thermal activated flux motion with a current dependence of the activation barrier [43]. Experiments are under way to systematically study the confinement effect on phase transition of vortex matter.

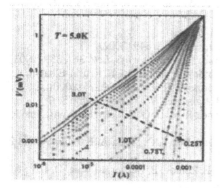

Figure 4. Current-voltage (I-V) characteristics of a 700 nm NbSe$_2$ nanowire in various magnetic fields obtained through four-probe transport measurements. A sign change in the curvature (in log-log plot) occurs in going from 0.75 T to 1.0 T and may be related to a second-order transition of the vortex matter.

CONCLUSIONS

In summary, we have successfully converted NbSe$_3$ nanowires and nanoribbons into superconducting NbSe$_2$ nanostructures through a controlled Se-reduction process. The conversion takes place in a narrow reaction temperature range of 600 °C ~ 700 °C. Both magnetization and transport measurements confirm the superconductivity of the converted NbSe$_2$ nanostructures, which have critical temperatures approaching the bulk value of 7.2 K. Confinement induced enhancement of the critical fields was observed in the synthesized NbSe$_2$ superconducting nanostructures. Remarkably, contrary to bulk crystals, the 'peak effect' does not appear in our NbSe$_2$ nanostructures.

ACKNOWLEDGMENTS

This work was supported by the US Department of Energy, BES-Materials Science, contract no. W-31-109-ENG-38. The scanning/transmission electron microscopy studies were performed in the Electron Microscopy Center at Argonne National Laboratory.

REFERENCES

1. J. T. Hu, T. W. Odom, and Ch. M. Lieber, *Acc. Chem. Res.* **32**, 435 (1999).
2. M. S. Dresselhaus, Y. M. Lin, O. Rabin, A. Jorio, A. G. Souza, M. A. Pimenta, R. Saito, G. G. Samsonidze, and G. Dresselhaus, *Mater. Sci. Eng. C* **23**, 129 (2003).
3. Z. L. Wang, *J. Phys.-Cond. Mat.* **16**, R829-R858 (2004).
4. Y. N. Xia, P. D. Yang , Y. G. Sun, Y. Y. Wu, B. Mayers, B. Gates, Y. D. Yin, F. Kim, and Y.

Q. Yan, *Adv. Mater.* **15**, 353 (2003).

5. X. F. Duan, Y. Huang, Y. Cui, J. F. Wang, and C. M. Lieber, *Nature* **409**, 66 (2001).

6. Z. W. Pan, Z. R. Dai, and Z. L. Wang, Science **291**, 1947 (2001).

7. J. H. Song, Y. Y. Wu, B. Messer, H. Kind, and P. D. Yang, *J. Am. Chem Soc.* **123**, 10397 (2001).

8. C. J. Murphy and N.R. Jana, *Adv. Mater.* **14**, 80 (2002).

9. G. Gu, B. Zheng, W. Q. Han, S. Roth, and J. Liu, *Nano Lett.* **2**, 849 (2002).

10. J. Q. Hu, Y. Jiang, X. M. Meng, C. S. Lee, and S. T. Lee, *Chem. Phys. Lett.* **367**, 339 (2003).

11. T. Thurn-Albrecht, J. Schotter, C. A. Kastle, N. Emley, T. Shibauchi, L. Krusin-Elbaum, K. Guarini, C. T. Black, M. T. Tuominen, and T. P. Russell, *Science* **290**, 2126 (2000).

12. L. Mohaddes-Ardabili, H. Zheng, S. B. Ogale, B. Hannoyer, W. Tian, J. Wang, S. E. Lofland, S. R. Shinde, T. Zhao, Y. Jia, L. Salamanca-Riba, D. G. Schlom, M. Wuttig, and R. Ramesh, *Nature Mater.* **3**, 533 (2004).

13. A. K. Geim, I. V. Grigorieva, S. V. Dubonos, J. G. S. Lok, J. C. Maan, A. E. Filippov, and F. M. Peeters, *Nature* **390**, 259 (1997).

14. C. A. Bolle, V. Aksyuk, F. Pardo, P. L. Gammel, E. Zeldov, E. Bucher, R. Boie, D. J. Bishop, and D. R. Nelson, *Nature* **399**, 43 (2000).

15. A. K. Geim, S. V. Dubonos, I. V. Grigorieva, K. S. Novoselov, F. M. Peeters, and V. A. Schweigert, *Nature* **407**, 55 (2000).

16. L. F. Chibotaru, A. Ceulemans, V. Bruyndoncx, and V. V. Moshchalkov, *Nature* **408**, 833 (2000).

17. A. Bezryadin, C. N. Lau, and M. Tinkham, *Nature* **404**, 971 (2000).

18. A. Rogachev and A. Bezryadin, *Appl. Phys. Lett.* **83**, 512 (2003).

19. A. Rogachev, A. T. Bollinger, and A. Bezryadin, *Phys. Rev. Lett.* **94**, 017004 (2005).

20. S Michotte, S. Matefi-Tempfli, and L. Piraux, *Appl. Phys. Lett.* **82**, 4119 (2003).

21. C. Y. Han, Z. L. Xiao, H. H. Wang, G. A. Willing, U. Geiser, U. Welp, W. K. Kwok, S. D. Bader, and G. W. Crabtree, *Plating and Surface Finishing* **91**, 40 (2004).

22. Y. L. Wang, X. C. Jiang, T. Herricks, and Y. N. Xia, *J. Phys. Chem. B* **108**, 8631 (2004).

23. Z. L. Xiao C. Y. Han, W. K. Kwok, H. W. Wang, U. Welp, J. Wang, and G. W. Crabtree, *J. Am. Chem. Soc.* **126**, 2316 (2004).

24. Q. Yang, J. Sha, X. Y. Ma, Y. J. Ji, and D. R. Yang, *Supercond. Sci. Technol.* **17**, L31 (2004).

25. Y. Y. Wu, B. Messer, and P. D. Yang, *Adv. Mater.* **13**, 1487 (2001).

26. Z. L. Xiao, E. Y. Andrei, and M. J. Higgins, *Phys. Rev. Lett.* **83**, 1664 (1999).

27. Z. L. Xiao, E. Y. Andrei, P. Shuk, and M. Greenblatt, *Phys. Rev. Lett.* **86**, 2431 (2001).

28. S. van Smaalen, J. L. de Boer, A. Meetsma, H. Graagsma, H-S. Sheu, A. Darovskikh, and P. Coppens, *Phys. Rev. B* **45**, 3103 (1992).

29. A. Prodan, N. Jug, H. J. P. van Midden, H. Bohm, F. W. Boswell, and J. C. Bennett, *Phys. Rev. B* **64**, 115423 (2001).

30. F. Levy and H. Berger, *J. Crystal Growth* **61**, 61(1983).

31. O. C. Mantel, F. Chalin, C. Dekker, H. S.J. van der Zant, Y. I. Latyshev, B. Pannetier, and P. Monceau, *Phys. Rev. Lett.* **84**, 538 (2000).

32. S. V. Zaitsev-Zotov, Microelectronic Eng. **69**, 549 (2003).

33. Y. S. Hor, Z. L. Xiao, U. Welp, Y. Ito, J. F. Mitchell, W. K. Kwok, and G. W. Crabtree, *Nano Lett.* **5**, 397 (2005).

34. J. McCarten, M. Maher, T. L. Adelman and R. E. Thorne, *Phys. Rev. Lett.* **63**, 2841(1989).

35. J. L. Hodeau, M. Marezio, C. Roucau, R. Ayroles, A. Meerschaut, J. Rouxel, and P. Monceau, *J. Phys. C: Solid State Phys.* **11**, 4117 (1978).
36. Y. S Hor, U. Welp, Y. Ito, Z. L. Xiao, J. F. Mitchell, W. K. Kwok, and G. W. Crabtree, *Appl. Phys. Lett.* **87**, 142506 (2005).
37. M. J. Higgins and S. Bhattacharya, *Physica C* **257**, 232 (1996).
38. S. Dubois, A. Michel, J. P. Eymery, J. L. Duvail, and L. Piraux, *J. Mater. Res.* **14**, 665 (1999).
39. Y. Paltiel, E. Zeldov, Y. Myasoedov, M. L. Rappaport, G. Jung, S. Bhattacharya, M. J. Higgins, Z. L. Xiao, E. Y. Andrei, P. L. Gammel, and D. J. Bishop, *Phys. Rev. Lett.* **85**, 3712 (2005).
40. R. H. Koch, V. Foglietti, W. J. Gallagher, G. Koren, A. Gupta and M. P. A. Fisher, *Phys. Rev. Lett.* **63**, 1511 (1989).
41. M. P. A. Fisher, *Phys. Rev. Lett.* **62**, 1415 (1989).
42. G. Blatter, M. V. Feigelman, V. B. Geshenbein, A. I. Larkin, and V. M. Vinokur, *Rev. Mod. Phys.* **66**, 1125 (1994).
43. Z. L. Xiao, J. Haring, C. Heinzel, and P. Ziemann, *Solid State Comm.* **95**, 153 (1995).

Mater. Res. Soc. Symp. Proc. Vol. 887 © 2006 Materials Research Society   0887-Q07-04

Fe-Core/Au-Shell Nanoparticles: Growth Mechanisms, Oxidation and Aging Effects

Kai Liu[1*], Sung-Jin Cho[2], Susan M. Kauzlarich[2†], J. C. Idrobo[1,3], Joseph E. Davies[1], Justin Olamit[1], N. D. Browning[4], Ahmed M. Shahin[5], Gary J. Long[5§], and Fernande Grandjean[6]

[1]Department of Physics, University of California, Davis, CA 95616, USA
[2]Department of Chemistry, University of California, Davis, CA 95616, USA
[3]Department of Physics, University of Illinois at Chicago, Chicago, IL 60607, USA
[4]Department of Chemical Engineering and Materials Science, University of California, Davis, CA 95616, USA
[5]Department of Chemistry, University of Missouri-Rolla, Rolla, MO 65409-0010, USA
[6]Department of Physics, University of Liège, B5, B-4000 Sart-Tilman, Belgium

ABSTRACT

We report the chemical synthesis of Fe-core/Au-shell nanoparticles (Fe/Au) by a reverse micelle method, and the investigation of their growth mechanisms and oxidation-resistant characteristics. The core-shell structure and the presence of the Fe and Au phases have been confirmed by transmission electron microscopy, energy dispersive spectroscopy, x-ray diffraction, Mössbauer spectroscopy, and inductively coupled plasma techniques. Additionally, atomic-resolution Z-contrast imaging and electron energy loss spectroscopy in a scanning transmission electron microscope have been used to study details of the growth processes. The Au-shells grow by nucleating on the Fe-core surfaces before coalescing. First-order reversal curves, along with the major hysteresis loops of the Fe/Au nanoparticles have been measured as a function of time in order to investigate the evolution of their magnetic properties. The magnetic moments of such nanoparticles, in the loose powder form, decrease over time due to oxidation. The less than ideal oxidation-resistance of the Au shell may have been caused by the rough Au surfaces. In a small fraction of the particles, off-centered Fe cores have been observed, which are more susceptible to oxidation. However, in the pressed pellet form, electrical transport measurements show that the particles are fairly stable, as the resistance and magnetoresistance of the pellet do not change appreciably over time. Our results demonstrate the complexity involved in the synthesis and properties of these heterostructured nanoparticles.

INTRODUCTION

Nanoparticles often exhibit novel properties as their physical dimensions become comparable to length scales in the nanometer range.[1-4] In particular, core/shell structured nanoparticles, due to the close proximity of the two functionally-different components, can exhibit enhanced properties and new functionality. Such structures not only are ideal for studying proximity effects, but are also suitable for structure stabilization as the shell layer protects the core from oxidation and corrosion. Additionally, the shell layer provides a platform for functionalization, such as coupling the core through the shell onto organic or other surfaces, thus making them potentially bio-compatible. Such core/shell structured magnetic nanoparticles are currently of interest for a wide variety of applications, for example, biological applications as magnetic resonance imaging (MRI) agents,[5] cell tagging and sorting,[6-9] hyperthermia treatment,[6, 10, 11] and targeted drug delivery.[12]

In these areas of research, the core/shell structure, particle size, shape and surface properties are important. In particular, iron oxides such as Fe_2O_3, Fe_3O_4, and MFe_2O_4 (M=Fe, Co, Mn) can be prepared as monodispersed surface derivatized nanoparticles.[13-17] Progress has also been made with the production of Co and Fe nanoparticles, as well as nanorods prepared by solution methods.[18-21] In spite of these advancements and exciting attributes, core/shell structured nanoparticles present enormous synthetic challenges, particularly in terms of the growth mechanisms of the core/shell structures and independent control over core/shell dimensions. For example, in Au coated Fe nanoparticles[22-26] prepared by a reverse micelle method, recent x-ray absorption spectroscopy studies have shown that such particles often consist of oxidized Fe cores.[23, 24] The saturation magnetization values are usually much smaller than that of the bulk Fe.[27] The long term integrity of the core/shell structure casts doubts on the usefulness of such nanoparticles. It was proposed that there may be grain boundaries in Au shells that allow for diffusion of oxygen and oxidation of the metallic cores.[23] An alternate explanation was that the Fe cores may not be centered in the micelles, resulting in asymmetric Au shells.[24]

There are a few key issues to be addressed: 1) are the as-made Fe cores metallic or oxidized? 2) when and how does the oxidation of the Fe cores occur, if any? Here we report a series of studies on Fe-core/Au-shell nanoparticles to answer these questions. Specifically, we investigate the physical properties of these nanoparticles both as a function of time (as made vs. over time) and under different conditions (annealed, stored in loose powder vs. pressed pellet form).

SYNTHESIS

Fe/Au nanoparticles were synthesized as previously reported.[27-30] The reaction was carried out in a reverse micelle reaction under argon gas by utilizing Schlenk line anaerobic techniques. Cetyltrimethylammonium bromide (CTAB) was used as the surfactant, octane as the oil phase, and 1-butanol as the co-surfactant. The water droplet size of the reverse micelle was controlled by the molar ratio of water to surfactant. Iron nanoparticles were prepared by the reduction of Fe^{2+} (in 1.2 mmol of $FeSO_4$) with $NaBH_4$ (2.4 mmol). The mixture was stirred at room temperature for 1 hour. The dark powder was separated from the solvent with a magnet and washed with CH_3OH twice and dried under vacuum. To create an Au shell on the Fe core, 0.8 mmol of $HAuCl_4$ was prepared as a micelle solution and added to the solution of $FeSO_4$ and $NaBH_4$. An additional 2.9 mmol of $NaBH_4$ micelle solution was immediately added to the solution that was subsequently left stirring at room temperature overnight. A dark precipitate was separated with a magnet and washed with CH_3OH twice to remove any nonmagnetic particles and organic surfactant. The sample was dried in vacuum.

STRUCTURAL AND CHEMICAL CHARACTERIZATIONS

X-ray diffraction (XRD) measurements were performed on a Scintag PAD-V diffractometer with Cu K_α radiation at a wavelength of 1.5406 Å. The XRD patterns of Fe/Au nanoparticles show peaks that can be assigned to both α-Fe and Au, consistent with previous reports[22] (Fig. 1). The Au and Fe diffraction peaks overlap and unambiguous evidence for the presence of Fe is not possible from XRD alone. An additional high-resolution x-ray diffraction pattern was collected on beam line 2-1 at the Stanford Synchrotron Radiation Laboratory. The pattern obtained with synchrotron radiation is very similar to that obtained with the conventional diffractometer, except for additional Au and α-Fe peaks due to the shorter wavelength and a wider 2θ range.

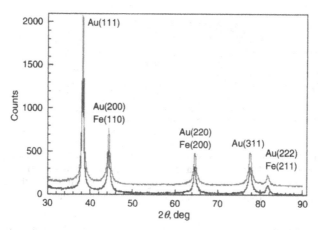

Figure 1. X-ray diffraction patterns of Au-coated Fe nanoparticles obtained with a Cu K_α radiation immediately after synthesis (lower blue curve) and one week after synthesis (upper green curve).

To investigate whether or not there is amorphous Fe or Fe-oxide present in the sample, the Fe/Au product was annealed in air at 400 °C overnight. Any amorphous Fe will oxidize and crystallize and any Fe-oxide present should become crystalline and be detectable by XRD. However, no new peak is observed, suggesting that there is no amorphous species present at room temperature. The crystallite size, calculated from the Au (111) reflection using the Scherrer formula and calibrated for instrumentation width, is 19 nm. To determine whether or not any Fe-oxide forms in the sample upon aging, the Fe/Au nanoparticles were stored in air for one week and the powder diffraction pattern was remeasured. The diffraction pattern again is similar to the original pattern and reveals no new diffraction peaks (Fig. 1).

Transmission electron microscopy (TEM) studies of the nanoparticles were obtained on a Philips CM-12 transmission electron microscope. Figure 2a shows a TEM image of the nanoparticles obtained from this synthesis, with a size of about 18±4 nm, consistent with the XRD results. Energy dispersive x-ray spectroscopy confirms the presence of both Fe and Au.

To further investigate how the Au shell is formed on these nanoparticles, we have used the high resolution Z-contrast imaging capability of scanning transmission electron microscopy (STEM). A typical image is shown in Fig. 2b, exhibiting a darker region (lower contrast) located at the center of the nanoparticle, surrounded by a brighter region. The pronounced contrast difference indicates the difference in chemical composition within the nanoparticles. Au, as a heavy element, scatters electrons more strongly than Fe, which has a smaller atomic number. Consequently, in the Z-contrast image shown in Fig. 2b, the brighter regions within the nanoparticle are Au rich while the darker regions are Fe rich and Au poor. Change of contrast can also be produced by change of thickness within the nanoparticle. However, electron energy loss spectroscopy (EELS) data taken on the two different regions do not show change in the background signal, indicating that the thickness is constant within the nanoparticle. Thus, this change of contrast is a strong indication that the nanoparticle is composed of a core Fe phase coated by Au.

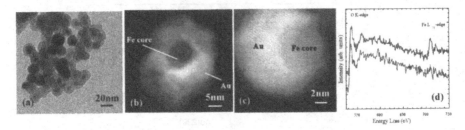

Figure 2. (a) Transmission electron microscopy image of Au-coated Fe nanoparticles. (b, c) Z-Contrast image of an Au-coated Fe nanoparticle obtained by scanning transmission electron microscopy. (d) Oxygen K-edge and Fe L_{23}-edge spectra acquired from the core (top, green solid curve) and surface (bottom, blue dotted curve) of the Fe/Au nanoparticle.

Also can be seen in the STEM image shown in Fig. 2b, the Au coating is continuous, but shows topographical roughness on the nanometer scale. These results would be consistent with the hypothesis that the Au-shell grows by nucleating from small nanoparticles on the Fe-core surface. This hypothesis of nanoparticle nucleation for shell formation on a core has been proposed by Pham et al.[31] for Au shell formation on silica nanoparticles. They have found that more complete surface capping can be accomplished by adding chemical reagents to help direct the shell growth. The rough surface could compromise the oxidation-resistance of the Au shell.

In a small fraction of the Fe/Au nanoparticles, off-centered Fe cores are observed, as shown in Fig. 2c. The Fe cores in such asymmetric nanoparticles are more likely to be exposed and oxidized, in agreement with the mechanism proposed by Ravel et al.[24] However, in the nanoparticles studied herein, such asymmetric nanoparticles are a minor occurrence.

To further investigate the chemical composition of the nanoparticles, atomic-resolution electron energy loss spectra were acquired. Figure 2d shows the oxygen (O) K-edge and Fe L_{23}-edge spectra from core (top solid curve) and surface (bottom dotted curve) of a Fe/Au nanoparticle. Each spectrum is the sum of 8 individual spectra with an acquisition time of 10 seconds and an energy resolution of 3 eV. An energy dispersion of 1 eV/pixel was used. The O K-edge onset was determined to be at 532 ± 1 eV.

As shown in Fig. 2d, the Fe signal is strongly present at the core spectrum. The spectrum of the surface of the particle shows only a trace signal for Fe, however the signal is slightly above the noise level. The Fe signal at the edge of the nanoparticle is likely coming from a residual Fe oxide phase around the Fe/Au nanoparticle as a result of the synthesis process. This may arise from inadequate rinsing of the nanoparticle or be due to Fe that does not get coated with Au and slowly oxidizes over time. This signal is low enough so as not to change the results of the analysis described below. To characterize the Fe oxidation state of the core, the L_3/L_2 white-line ratio was calculated. White-lines arise mainly from dipole selection rules due to transitions from the inner shell electrons to unoccupied states in the valence band.[32] The L_3/L_2 ratio was measured by the second derivative method, which has proven to characterize effectively Fe oxidation states.[31] The maximum of the two peaks on Fe core spectrum are located at 709 eV and 722 eV, for L_3 and L_2, respectively. The L_3/L_2 ratio measured on Fe core spectrum was 3.3 ± 0.8, characteristic of a metallic Fe phase. Oxygen signal was found in both spectra as shown in Fig. 2d, which comes mainly from oxygen on the silica support.

MAGNETIC PROPERTIES

Magnetic measurements were performed using a Quantum Design superconducting quantum interference device (SQUID) magnetometer. Magnetic hysteresis loops of the Fe/Au nanoparticles are shown in Fig. 3a. At 5 K, the particles display a coercivity of 400 Oe, a remanent magnetization of 14 emu/g, and a saturation magnetization M_S of 43 emu/g. Correcting for the composition of the nanoparticles, 26.5 at.% of Fe as determined from inductively coupled plasma (ICP) analysis, the saturation magnetization is 162 emu/(g-Fe), close to the expected value of 220 emu/g for bulk Fe. At 300 K, the Fe/Au nanoparticles still exhibit significant saturation magnetization, about 2/3 of the 5 K M_S, although the hysteresis has diminished. These results suggest that some Fe cores are large enough to behave like bulk Fe at room temperature.

To clarify the time scale of oxidation of Fe/Au nanoparticles, the saturation magnetization M_S was measured daily since right after synthesis. The nanoparticles were directly exposed to air, stored and measured in gel capsules during this study. After 5 days, M_S has decreased to 50 % of its initial value (Fig. 3b).This reduction suggests that there has been some oxidation of the core as the magnetization of Fe-oxides are lower than that of α-Fe.[33]

To further investigate the subtle changes of magnetic properties over time, first-order reversal curve (FORC) measurements were performed by using a Princeton Measurements vibrating sample magnetometer with a liquid helium continuous flow cryostat. For this study, 10 mg of Fe/Au nanoparticles were first dispersed in hexane under sonication inside a nitrogen-filled glove box, and then mixed with rubber cement. In this technique, a few hundred first-order reversal curves (FORC's) were measured in the following manner. After saturation, the magnetization, M, is measured with increasing applied field, H, starting from the reversal field, H_R, back to positive saturation. A family of FORC's is measured at different H_R values with equal field spacing, thus filling the interior of the major hysteresis loop. The FORC distribution is defined by a mixed second order derivative, [34-38]

$$\rho(H_R, H) \equiv -\frac{1}{2}\frac{\partial^2 M(H_R, H)}{\partial H_R \partial H}.$$

(1)

A three-dimensional plot of the distribution, ρ, versus H and H_R, i.e., a FORC diagram, can then be used to probe the details of the magnetization reversal. Alternatively, ρ can be plotted as a function of the local coercivity, H_c, and the bias field, H_b, after a $H_b = (H + H_R)/2$ and $H_c = (H$

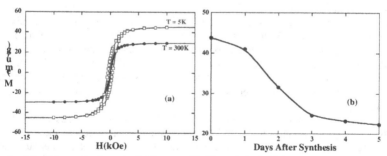

Figure 3. (a) Magnetic hysteresis loop at 5 K and 300 K. (b) Decay of saturation magnetization of exposed Fe/Au nanoparticles over time.

125

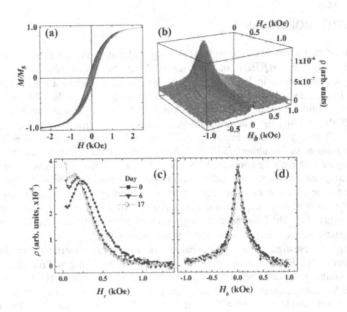

Figure 4. A family of first-order reversal curves (FORC) obtained at 35 K four days after the preparation of the Fe/Au nanoparticles, (a). The outer boundary delineates the major loop. The corresponding FORC distribution is shown in (b). The projection of the FORC distribution onto the H_c, (c), and H_b, (d), axes.

$- H_R)/2$ coordinate transformation.[34, 35] If a material is composed of a set of independent magnetic particles, the resulting FORC diagram will map the distribution of the coercivity and bias field of the collection of particles. For real systems, the FORC diagram also reveals any complex interactions that may occur among the particles, as will be illustrated below. Thus, the first-order reversal curves provide much more information than the ensemble average measured by typical magnetic major hysteresis loops. Details of the methodology and its applications have been described in prior publications.[34-38]

The first-order reversal curves, along with the major hysteresis loops, of a mixture of the Fe/Au nanoparticles with rubber cement have been measured as a function of time in order to monitor the evolution of their magnetic properties. These measurements were carried out after zero-field cooling to 35 K; a nominal field step of 0.02 kOe was used. The magnetization reversal curves filling the interior of the major loop are shown in Fig. 4a. The corresponding FORC distribution, ρ, is shown in terms of the coercivity, H_c, and the bias field, H_b, coordinates in Fig. 4b.

As is indicated in Fig. 4b, a non-zero FORC distribution ρ extends along the H_c axis, indicating a finite distribution in the local coercivity due to a finite particle size distribution. Furthermore, some subtle changes in the magnetic characteristics of the Fe/Au nanoparticles have been revealed by the FORC measurements. For example, a slow oxidation process of the Fe/Au nanoparticles over time is illustrated in Figs. 4c and 4d. The projection of the FORC distribution onto the H_c axis, see Fig. 4c, shows the coercivity distribution at the 0th, 6th and 17th day after synthesis. The average coercivity (peak position) is reduced with time, consistent with a gradual decrease in the Fe/Au core size due to oxidation.[39, 40] Furthermore, the coercivity distribution shown in Fig. 4c becomes narrower over time, indicating that particles with larger magnetic cores have experienced more oxidation and enhanced size-reduction.

It is also interesting to note the evolution of the distribution of the bias field, H_b, a distribution which is related to the average inter-particle spacing and the extent of their magnetic interactions.[38] The H_b distribution is obtained by projecting the FORC distribution onto the H_b axis, as is shown in Fig. 4d. If the Fe/Au nanoparticles are well dispersed, i.e., if the dipolar and exchange interactions between the particles are negligible, each particle would experience only the applied field and thus reverse its magnetization at its respective coercive field. The hysterons, or hysteresis loops for each particle, would then have zero bias, resulting in a FORC distribution that has a narrow ridge along H_c centered at $H_b = 0$. Because the Fe/Au nanoparticles are not fully dispersed, their interactions are manifested as a distribution of the bias field, H_b, as is shown in Fig. 4b. The bias field distribution changes negligibly with time for the Fe/Au nanoparticles, see Fig. 4d, indicating that, although the particles are undergoing oxidation, the average particle spacing and the interactions between the particles remain essentially unchanged.

For this sample of Fe/Au nanoparticles embedded in rubber cement, the major loop coercivity and saturation magnetization are also observed to decrease over time. At 35 K, the major loop coercivity decreases from 170 to 110 Oe over 17 days, with a decay constant of ~42 days required to reach $1/e$ or 37 %. The decreases in the coercivity and saturation magnetization further indicate sample oxidation over time. The decay constant is much longer than that observed if the sample was exposed directly to air. The rubber cement offers some degree of protection from the atmosphere, leading to a slower oxidation with time.[28, 29]

MÖSSBAUER SPECTRAL STUDIES

The Mössbauer spectra were measured at 78 K and 295 K on a conventional constant-acceleration spectrometer that utilized a room-temperature rhodium-matrix [57]Co source and was calibrated at room temperature with α-Fe foil. The studies were performed at these temperatures because of the requirement that the samples not be exposed to oxygen or moisture.[41] Immediately after synthesis the sample was placed and sealed in a nitrogen-filled double-layer vial; this vial was then shipped from University of California – Davis to the University of Missouri–Rolla by overnight express mail. The absorber, which was prepared and placed in the cryostat in a pure dry nitrogen atmosphere, contained ~20 mg/cm^2 of material finely dispersed in deoxygenated boron nitride.

The Mössbauer spectra of the Fe/Au nanoparticles are shown in Fig. 5. The spectra exhibit a sharp sextet, with a relative area of ca. 40 % and hyperfine parameters that are typical of crystalline α-Fe.[41] This verifies the production of α-Fe particles using the reverse-micelle reduction route. In addition, the spectra exhibit the presence of paramagnetic high-spin Fe(II), high-spin Fe(III), and a broad sextet.

The assignment of the high-spin Fe(II) doublet present in the spectra of Fig. 5 to a specific compound is difficult. Because FeSO$_4$ was used in the preparation, it is tempting to assign this doublet to a ferrous sulfate. However, the observed hyperfine parameters do not match either those of anhydrous FeSO$_4$ or those of FeSO$_4$·7H$_2$O. The hyperfine parameters of the high-spin Fe(III) doublet are typical of superparamagnetic particles of γ-Fe$_2$O$_3$ or Fe$_3$O$_4$;[42] it is not possible

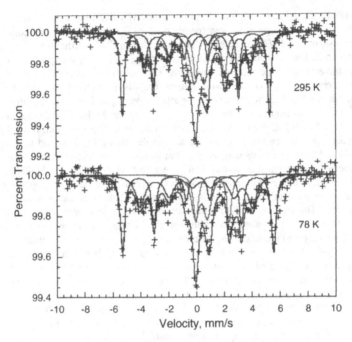

Figure 5. The Mössbauer spectra of the Fe/Au nanoparticles, obtained within a week of synthesis, protected from air-oxidation. The Fe(II), Fe(III), α-Fe, and $Fe_{0.73}B_{0.27}$ components are shown in green, red, black, and blue, respectively.

on the basis of the hyperfine parameters to differentiate these two oxides. This assignment agrees with the presence of γ-Fe_2O_3 in the Fe/Au nanoparticles[24] and the presence of Fe_3O_4 in the Fe-Au composite particles[23] observed by x-ray absorption spectroscopy. The Fe(III) doublet relative areas of ca. 22 % in the Fe/Au nanoparticles indicate that only a small fraction of the sample behaves as superparamagnetic particles on the ^{57}Fe Mössbauer-effect timescale of 10^{-8} s.

The broad sextet with a hyperfine field of ca. 25 T was initially assigned to Fe-Au alloys.[28] However, in view of its presence in the Mössbauer spectra of the Fe nanoparticles,[30] this assignment is revised. The presence of boron revealed by the inductively coupled plasma elemental analysis, as well as the observation of Fe_2B in the Fe nanoparticles prepared via a similar route by Glavee et al.,[43] support the presence of Fe_2B or $Fe_{1-x}B_x$ in the Fe/Au nanoparticles. The hyperfine parameters of the broad sextet agree reasonably well with those of amorphous $Fe_{1-x}B_x$ alloys,[44, 45] prepared by a method similar to that used herein. The 295 K hyperfine field of 23.4 ± 0.4 T observed herein and the linear dependence[45] of the field with the boron content yield an x value of 0.27 ± 0.02 and an average composition of $Fe_{0.73}B_{0.27}$. However, we cannot rule out the presence of Fe-Au alloys in the Fe/Au nanoparticles and their contribution to the broad sextet.

ELECTRICAL TRANSPORT PROPERTIES

Electrical transport properties have been measured in pressed pellets of these nanoparticles. Pellets were prepared by cold-pressing nanoparticles into a 6 mm die under a 2×10^7 Pa pressure for 10 minutes. Electrical leads were attached by silver paint onto the pellet. The temperature-dependence of resistance and magnetoresistance at 5 K were measured repeatedly over 2 months to monitor the time scale of Fe oxidation.

Three pellets were prepared in the following way. Pellet **A** was pressed from as-made nanoparticles immediately after synthesis and measured right away; the remaining nanoparticles were stored as loose powders and exposed to air for one month, and then pressed to form pellet **B**; pellet **C** was prepared from as-made particles and stored in air in the pellet form; the electrical transport properties were measured over time for pellet **C**. The results are shown in Fig. 6.

For pellet **A**, right after synthesis, the resistance decreases slightly with decreasing temperature, as shown in Fig. 6a. This positive temperature coefficient of resistance is a signature of metallic conduction, in contrast to the negative temperature coefficient and thermally activated behavior seen in pellets of iron oxide nanoparticles.[33, 46] Thus the core-shell samples right after synthesis are still metallic.

At 5 K, the resistance decreases in a magnetic field, resulting in a negative magnetoresistance (MR). The MR, defined as $[R(H) - R(0)]/R(0) = \Delta R/R(0)$, is about -0.2% in a field of 40 kOe (Fig. 6a inset). The negative MR is the giant magnetoresistance effect caused by spin dependent scattering, similar to those seen in magnetic granular solids.[47] In the pellet of Fe/Au nanoparticles, the Fe cores serve as magnetic scattering centers. At low fields the magnetic moments of the Fe cores are random, resulting in a spin-disordered high resistance state. The application of a magnetic field helps to align the Fe core moments and reduces the spin-disorder. This in turn reduces the spin-dependent scattering and leads to a low resistance state, hence the negative MR.

For pellet **B** (after month-long exposure to air), the resistance values are much larger, accompanied by a negative temperature coefficient (from 10 MΩ at 300 K to 26 GΩ at 150 K,

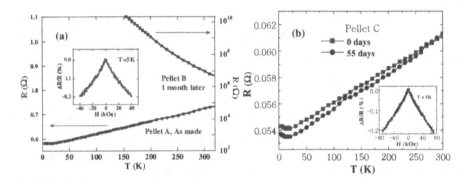

Figure 6. Temperature dependence of resistance in zero magnetic field of (a) pellet **A** (black solid circles) and pellet **B** (blue open squares), and (b) pellet **C**. Inset shows the magnetic field dependence of magnetoresistance at 5 K of (a) pellet **A** and (b) pellet **C**.

measured by a Keithley 617 electrometer with a 200 GΩ impedance) and a thermally activated behavior at high temperatures (Fig. 6a). Clearly, the nanoparticles left in air have oxidized.

For pellet C, the temperature dependence of the resistance of a pellet is shown in Fig. 6b. The resistance decreases slightly with decreasing temperature. Furthermore, MR has been measured at 5 K, as shown in the inset of Fig. 6b. A similar negative MR effect was observed, confirming the presence of magnetic scattering centers. These electrical measurements have been repeated many times over a 55-day period. The results obtained are always the same as those obtained right after synthesis. We note that the resistivity measurement is susceptible to a percolated conduction path through Au, thus less sensitive to Fe oxidation. In contrast, the MR effect is sensitive to Fe oxidation as it is due to spin-dependent scattering at the interfaces between Au and Fe as well as within the magnetic Fe cores. Any oxidation of the Fe cores, into magnetic or non-magnetic Fe-oxides, will change this spin-dependent scattering process and result in a change in MR. The lack of appreciable changes in both resistivity and MR results demonstrates that when pressed into a pellet, although still exposed to air, the Fe/Au nanoparticles are stable over time.

SUMMARY

We have synthesized Fe-core/Au-shell nanoparticles by a reverse micelle method, and investigated their growth mechanisms and oxidation-resistant characteristics. The core/shell heterostructure and the presence of the Fe and Au phases have been clearly confirmed. Our experiments demonstrate that metallic Fe cores are indeed formed after synthesis. The Au shells do not protect the Fe cores completely from oxidation. Depending on the level of exposure to air, the Fe cores oxidize over time at different rates. Additionally, an amorphous $Fe_{1-x}B_x$ or FeAu phase exist. The Au shells appear to grow by nucleating at selected sites on the Fe core surfaces before coalescing. The rough surfaces could compromise the oxidation-resistance of the Au shells. A small fraction of the particles have off-centered Fe cores, which are more susceptible to oxidation. In the pressed pellet form, electrical transport measurements show that the particles are fairly stable, as the resistance and magnetoresistance of the pellet do not change appreciably over time.

The ensemble of results presented herein suggests that a number of detailed analyses with different techniques probing different components and properties are required to fully understand any complex nanoparticles. These results also suggest that further efforts for synthetic optimization of Fe/Au nanoparticles are warranted.

ACKNOWLEDGEMENTS

We thank Hsiang-Wei Chiu for obtaining the synchrotron x-ray diffraction data, John Neil for technical support during the analysis of the x-ray diffraction results. This research was supported by the National Science Foundation (DMR-0120990, CHE-0210807, and ECS-0508527), the American Chemical Society (PRF-43637-AC10), the Alfred P. Sloan Foundation, and the University of California (CLE). Portions of this research were carried out at the Stanford Synchrotron Radiation Laboratory, a national user facility operated by Stanford University on behalf of the U.S. Department of Energy, Office of Basic Energy Sciences. F. G. acknowledges with thanks the financial support of the Fonds National de la Recherche Scientifique, Belgium, through grant 9.456595 and the Ministère de la Région Wallonne for grant RW/115012.

REFERENCES

*   Electronic address: kailiu@ucdavis.edu.

†   Electronic address: smkauzlarich@ucdavis.edu.

§   Electronic address: glong@umr.edu.

[1]   D. D. Awschalom and S. von Molnár, in *Nanotechnology (Chapter 12)*, edited by G. Timp (Springer-Verlag, New York, 1998).

[2]   K. Ounadjela and R. L. Stamps, in *Handbook of Nanostructured Materials and Nanotechnology (Chapter 9)*, edited by H. S. Nalwa (Academic Press, San Diego, 2000), Vol. 2.

[3]   C. Ross, An. Rev. Mater. Res. **31**, 203 (2001).

[4]   J. I. Martin, J. Nogues, K. Liu, J. L. Vicent, and I. K. Schuller, J. Magn. Magn. Mater. **256**, 449 (2003).

[5]   D. K. Kim, Y. Zhang, J. Kehr, T. Klason, B. Bjelke, and M. Muhammed, J. Magn. Magn. Mater. **225**, 256 (2001).

[6]   C. M. Niemeyer, Angewandte Chemie, Int. Ed. **40**, 4128 (2001).

[7]   G. X. Li and S. X. Wang, IEEE Trans. Magn. **39**, 3313 (2003).

[8]   G. X. Li, S. X. Wang, and S. H. Sun, IEEE Trans. Magn. **40**, 3000 (2004).

[9]   A. R. Bausch, W. Moller, and E. Sackmann, Biophys. J. **76**, 573 (1999).

[10]   S. Mornet, S. Vasseur, F. Grasset, and E. Duguet, J. Mater. Chem. **14**, 2161 (2004).

[11]   P. Gangopadhyay, S. Gallet, E. Franz, A. Persoons, and T. Verbiest, IEEE Trans. Magn. **41**, 4194 (2005).

[12]   M. Zahn, J. Nanopar. Res. **3**, 73 (2001).

[13]   C. J. O'Connor, C. Seip, C. Sangregorio, E. Carpenter, S. Li, G. Irvin, and V. T. John, Mole. Crys. Liq. Crys. Sci. Tech. A **335**, 1135 (1999).

[14]   D. Wang, J. He, N. Rosenzweig, and Z. Rosenzweig, Nano Lett. **4**, 409 (2004).

[15]   S. Sun and H. Zeng, J. Am. Chem. Soc. **124**, 8204 (2002).

[16]   L. T. Kuhn, A. Bojesen, L. Timmermann, M. M. Nielsen, and S. Morup, J. Phys.: Cond. Mat. **14**, 13551 (2002).

[17]   S. H. Sun, H. Zeng, D. B. Robinson, S. Raoux, P. M. Rice, S. X. Wang, and G. X. Li, J. Am. Chem. Soc. **126**, 273 (2004).

[18]   V. F. Puntes, K. M. Krishnan, and A. P. Alivisatos, Science **291**, 2115 (2001).

[19]   S.-J. Park, S. Kim, S. Lee, Z. G. Khim, K. Char, and T. Hyeon, J. Am. Chem. Soc. **122**, 8581 (2000).

[20]   F. Dumestre, B. Chaudret, C. Amiens, P. Renaud, and P. Fejes, Science **303**, 821 (2004).

[21]   J. Bai and J.-P. Wang, Appl. Phys. Lett. **87**, 152502 (2005).

[22]   E. E. Carpenter, C. Sangregorio, and C. J. O'Connor, IEEE Trans. Magn. **35**, 3496 (1999).

[23]   T. Kinoshita, S. Seino, K. Okitsu, T. Nakayama, T. Nakagawa, and T. A. Yamamoto, J. Alloy. Comp. **359**, 46 (2003).

[24]   B. Ravel, E. E. Carpenter, and V. G. Harris, J. Appl. Phys. **91**, 8195 (2002).

[25]   E. E. Carpenter, J. Magn. Magn. Mater. **225**, 17 (2001).

[26]   C. J. O'Connor, V. Kolesnichenko, E. Carpenter, C. Sangregorio, W. Zhou, A. Kumbhar, J. Sims, and F. Agnoli, Synth. Met. **122**, 547 (2001).

[27]   J. Lin, W. Zhou, A. Kumbhar, J. Wiemann, J. Fang, E. E. Carpenter, and C. J. O'Connor, J. Solid St. Chem. **159**, 26 (2001).

[28] S.-J. Cho, S. M. Kauzlarich, J. Olamit, K. Liu, F. Grandjean, L. Rebbouh, and G. J. Long, J. Appl. Phys. **95**, 6804 (2004).

[29] S.-J. Cho, J.-C. Idrobo, J. Olamit, K. Liu, N. D. Browning, and S. M. Kauzlarich, Chem. Mater. **17**, 3181 (2005).

[30] S.-J. Cho, A. M. Shahin, G. J. Long, J. E. Davies, K. Liu, F. Grandjean, and S. M. Kauzlarich, Chem. Mater., in press (2006); cond-mat/0512413.

[31] T. Pham, J. B. Jackson, N. J. Halas, and T. R. Lee, Langmuir **18**, 4915 (2002).

[32] R. F. Egerton, *Electron Energy-Loss Spectroscopy in The Electron Microscope*, 1986).

[33] K. Liu, L. Zhao, P. Klavins, F. E. Osterloh, and H. Hiramatsu, J. Appl. Phys. **93**, 7951 (2003).

[34] C. R. Pike, A. Roberts, and K. L. Verosub, J. Appl. Phys **85**, 6660 (1999).

[35] H. G. Katzgraber, F. Pázmándi, C. R. Pike, K. Liu, R. T. Scalettar, K. L. Verosub, and G. T. Zimányi, Phys. Rev. Lett. **89**, 257202 (2002).

[36] J. E. Davies, O. Hellwig, E. E. Fullerton, G. Denbeaux, J. B. Kortright, and K. Liu, Phys. Rev. B **70**, 224434 (2004).

[37] J. E. Davies, O. Hellwig, E. E. Fullerton, J. S. Jiang, S. D. Bader, G. T. Zimanyi, and K. Liu, Appl. Phys. Lett. **86**, 262503 (2005).

[38] J. E. Davies, J. Wu, C. Leighton, and K. Liu, Phys. Rev. B **72**, 134419 (2005).

[39] B. D. Cullity, *Intorduction to magnetic materials* (Addison-Wesley Pub. Co., Reading, Mass., 1972).

[40] K. Liu and C. L. Chien, IEEE Trans. Magn. **34**, 1021 (1998).

[41] G. J. Long, D. Hautot, Q. A. Pankhurst, D. Vandormael, F. Grandjean, J. P. Gaspard, V. Briois, T. Hyeon, and K. S. Suslick, Phys. Rev. B **57**, 10716 (1998).

[42] A. A. Novakova, V. Y. Lanchinskaya, A. V. Volkov, T. S. Gendler, T. Y. Kiseleva, M. A. Moskvina, and S. B. Zezin, J. Magn. Magn. Mater. **258-259**, 354 (2003).

[43] G. N. Glavee, K. J. Klabunde, C. M. Sorensen, and G. C. Hadjipanayis, Inorg. Chem. **34**, 28 (1995).

[44] N. Duxin, O. Stephan, C. Petit, P. Bonville, C. Colliex, and M. P. Pileni, Chem. Mater. **9**, 2096 (1997).

[45] S. Linderoth and S. Mørup, J. Appl. Phys. **69**, 5256 (1991).

[46] L. Savini, E. Bonetti, L. Del Bianco, L. Pasquini, L. Signorini, M. Coisson, and V. Selvaggini, J. Magn. Magn. Mater. **262**, 56 (2003).

[47] J. Q. Xiao, J. S. Jiang, and C. L. Chien, Phys. Rev. Lett. **68**, 3749 (1992).

Mater. Res. Soc. Symp. Proc. Vol. 887 © 2006 Materials Research Society 0887-Q07-07

MAGNETIC AGING

Ralph Skomski, Jian Zhou, R. D. Kirby, and D. J. Sellmyer

Department of Physics and Astronomy and Center for Materials Research and Analysis, University of Nebraska, Lincoln, NE 68588

ABSTRACT

Thermally activated magnetization reversal is of great importance in areas such as permanent magnetism and magnetic recording. In spite of many decades of scientific research, the phenomenon of slow magnetization dynamics has remained partially controversial. It is now well-established that the main mechanism is thermally activated magnetization reversal, as contrasted to eddy currents and structural aging, but the identification of the involved energy barriers remains a challenge for many systems. Thermally activated slow magnetization processes proceed over energy barriers whose structure is determined by the micromagnetic free energy. This restricts the range of physically meaningful energy barriers. An analysis of the underlying micromagnetic free energy yields power-law dependences with exponents of 3/2 or 2 for physically reasonable models, in contrast to arbitrary exponents m and to $1/H$-type laws.

INTRODUCTION

Magnetic properties such as the remanent magnetization are weakly time-dependent. Depending on the context, this degradation is known as magnetic aging or magnetic viscosity [1-3]. For example, permanent magnets loose a small fraction of their remanence each decade [4], and the long-term stability of stored information is a major concern in ultrahigh-density magnetic recording media. Typical relaxation or equilibration times vary between less than a second in superparamagnetic particles and millions of years in magnetic rocks. Similar time-dependent magnetization processes are important soft magnets, although the involved time scales are often in milli- or nanosecond ranges.

Initially, magnetic aging was believed to reflect mechanisms such as eddy currents, but soon it became clear that eddy-current contributions are usually negligible. Some magnetization changes are due to structural aging, but most mechanisms reflect the ther-

magnetic rocks   permanent magnets   recording media

Figure 1. Systems where magnetic aging is important.

mal activation over magnetic energy barriers [1]. Unlike structural aging, magnetic aging is reversible, so that the application of a large positive magnetic field re-establishes the original magnetization state. A key question is the physical nature of the involved energy barriers E_a over which thermal activation occurs. Various partially exclusive field-dependences $E_a(H)$ have been proposed, and there is a continuing debate about the applicability of these expressions [5-7].

STRUCTURAL EFFECTS

There are two main types of magnetic aging, namely structural aging and thermally activated magnetization processes. Structural aging refers to both the crystal structure and to defects. Time-dependent structural changes are important, for example, in metastable intermetallics such as $Sm_2Fe_{17}N_3$ permanent magnets, where the decay into SmN and Fe limits the maximum application temperature. However, the magnetism related to the decay is relatively uninteresting, as compared to the anisotropy and Curie-temperature enhancement in the material [8].

A mechanism involving both structural and magnetic degrees of freedom is the Snoek aftereffect in pinning-type magnets. It occurs in steels and related materials and means that magnetic domain walls interact with diffusing carbon atoms. The *static* coercivity of pinning-type magnets is the reverse field necessary to overcome the pinning energy barrier, as indicated in Fig. 2(a). The Snoek effect, Fig. 2(b), means that the energy barrier is reduced by the diffusion of interstitial carbon or nitrogen atoms.

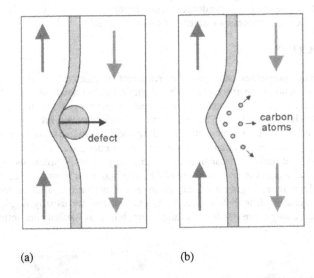

(a)   (b)

Figure 2. Magnetic aging and domain-wall motion: (a) ordinary structural defects and (b) Snoek effect.

THERMALLY ACTIVATED MAGNETIZATION REVERSAL

Figure 3 compares thermally activated magnetization reversal with the so-called 'static' magnetization reversal, where the micromagnetic energy barriers Ea are overcome due to a reverse magnetic field. The relaxation time obeys so-called Néel-Brown law

$$\tau = \tau_0 \exp\left(\frac{E_a}{k_B T}\right) \tag{1}$$

which goes back to the early decades of the 20th century [1]. Here τ is the relaxation time and $\tau_0 = 1/\Gamma_0$ is an inverse attempt frequency of order 10^{-10} s. Depending on the context, Eq. (1) is also known as the Néel or Néel-Brown relaxation law.

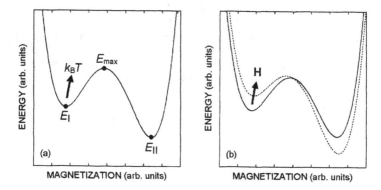

Fig. 3. Thermally activated magnetization reversal (left) and static coercivity (right).

Typically, thermal activation is a small correction to the leading field-dependence $E_a(H)$. Inverting Eq. (1) yields the energy barriers $E_a = k_B T \ln(\tau/\tau_0)$ that are thermally accessible after some time τ. Taking $\tau = 100$ s, we reproduce the famous result $E_a \approx 25$ $k_B T$. At room temperature, this corresponds to a temperature equivalent of about 7'500 K. However, energy barriers often exceed 100'000 K [3, 6], so that some supporting external field is necessary to reduce them to 25 $k_B T$ before thermal activation takes over.

Power Laws with m = 3/2 and m = 2

The starting point of the description of thermally activated magnetization reversal is the micromagnetic (free) energy $F(\mathbf{M}(\mathbf{r}), K_1(\mathbf{r}), A(\mathbf{r}))$, where $\mathbf{M}$, K_1 and A are the local magnetization, anisotropy, and exchange stiffness, respectively. Figure 3 is an example of a micromagnetic free energy, although only one degree of freedom, namely the relevant reversal mode, is shown in the figure. Series expansion of the micromagnetic free energy yields [6]

$$E_a \sim (H_0 - H)^m \tag{2}$$

where H_o is the static coercivity and the exponent $m = 3/2$ or $m = 2$ depends on the symmetry of the magnet [5, 6, 9]. Most systems have $m = 3/2$, but $m = 2$ for highly symmetric systems, such as aligned Stoner-Wohlfarth particles.

There have been approaches to treat m as an adjustable or field-dependent parameter, and it has been has argued that m implies some kind of averaging over energy barriers. Other proposed dependences are $m = 1$ laws and relations such as $1/H$ and $1/H - 1/H_o$ [7, 10]. The discussion has been fueled by the popular belief that the exponents $m = 2$ and $m = 3/2$ are limited to specialized or highly simplified models. In fact, they go far beyond the Stoner-Wohlfarth approach [5] and describe a broad range of coherent and incoherent magnetization processes [6]. It has also been possible to derive the $m = 2$ and $m = 3/2$ modes from realistic energy landscapes F. Other modes involve very crude approximations, are incompatible with the real structure of the magnets, or misinterpret the physics of the magnetization reversal. For example, the $1/H$ law corresponds to the physically unreasonable prediction of an infinite zero-temperature coercivity, whereas the $1/H - 1/H_o$ dependence reduces, by series expansion, to $m = 1$.

Linear laws, $m = 1$, are used in simplified models and to evaluate experimental data in terms of activation volumes V^*. However, so far it has not been possible to derive them from physically reasonable energy landscapes [6], and V^* is not necessarily equal to the physical switching volume V [4]. Figure 4 shows a fictitious pinning energy landscape that would yield a linear law. In reality, the singularities responsible for the (piecewise) linear nature of $E_a(H)$ are smoothed by the continuous domain-wall profile. The smoothing affects just a few nanometers, but the corresponding energies are typically larger than $25 k_B T$.

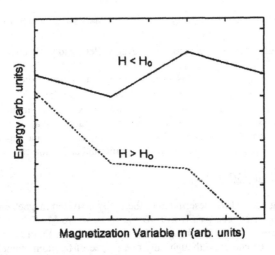

Fig. 4. Example of a fictitious energy landscape with $m = 1$.

Effect of Real Structure

Thermally excited modes are supported by real-structure features and correspond to static reversal modes. There is no justification for using arbitrary magnetization modes to fit experimental data. Such modes have often energies of the order of $\delta_B^3 K_1 \sim 100\,000$ K, as contrasted to the accessible range of 25 $k_B T$ [11]. Figure 5 shows some examples of physically meaningful (a) and physically unrealistic modes (b-c). Spin configurations such as Fig. 5(b) and (c) are limited to though frequently occur in *imperfect* magnets. The inhomogenity or randomness yields a renormalization of the zero-field energy and of H_o, but leaves the functional structure of Eq. (2) unchanged.

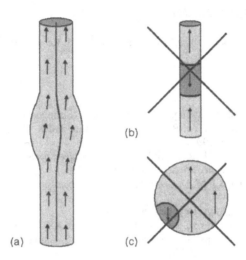

Figure 5. Physically meaningful (a) and arbitrary (b-c) magnetization modes. Note that (b) and (c) refer to homogeneous ellipsoids of revolution; defects drastically change the model predictions.

An exception is very small or 'superparamagnetic' particles, where $E_a \sim 25\ k_B T$ at room temperature [12]. This limit requires the inclusion of higher-order terms in Eq. (2), and there are corrections to the simple $m = 3/2$ and $m = 2$ power laws. However, even in this case, $E_a(H)$ derives from the micromagnetic free energy and must reproduce Eq. (2) in the limit of low temperatures. Otherwise, $E_a(H)$ acquires the character of a phenomenological fitting function that describes a relatively narrow energy or temperature window.

CONCLUSIONS

In summary, we have investigated the physical origin of magnetic aging. Emphasis has been on the energy barriers responsible for thermally activated magnetization reversal. To obtain meaningful expressions for the relaxation time, it is necessary to start from physically reasonable energy landscapes, based on the microstructure of the magnet. Neglecting superparamagnetic effects, the energy barriers responsible for thermally activated slow magnetization dynamics are of the power-law type, with exponents $m = 3/2$ or 2, depending on the symmetry of the problem. In contrast to widespread belief, these laws are not restricted to Stoner-Wohlfarth particles, but describe a broad range of pinning and nucleation mechanisms. Arbitrary exponents m and $1/H$-type energy-barrier dependences are not supported by the real structure of the magnets and yield physically unreasonable predictions, such as infinite zero-temperature coercivities.

ACKNOWLEDGEMENT

This research is supported by NSF-MRSEC, the W. M. Keck Foundation, INSIC, and CMRA.

REFERENCES

[1] R. Becker and W. Döring, *Ferromagnetismus* (Springer, Berlin, 1939).
[2] D. J. Sellmyer, M. Yu, R. A. Thomas, Y. Liu, and R. D. Kirby, Phys. Low-Dim. Struct. **1-2**, 155 (1998).
[3] D. Givord and M. F. Rossignol, in: *Rare-earth iron permanent magnets*, Ed.: J. M. D. Coey (University Press, Oxford, 1996) p. 218.
[4] R. Skomski and J. M. D. Coey, *Permanent Magnetism*, Institute of Physics, Bristol 1999.
[5] R. H. Victora, Phys. Rev. Lett. **63**, 457 (1989).
[6] R. Skomski, J. Phys.: Condens. Matter **15** (2003) R841.
[7] J. Moritz, B. Dieny, JP Nozières, Y. Pennec, J. Camarero, and S. Pizzini, Phys. Rev. B **71**, 100402 (2005).
[8] R. Skomski, in: *Rare-Earth—Iron Permanent Magnets*, Ed.: J. M. D. Coey, University Press, Oxford 1996, p. 178-217.
[9] L. Néel, J. de Phys. Rad. **11**, 49 (1950).
[10] T. Egami, Phys. Stat. Sol. (a) **20**, 157 (1973); (b) **57**, 211 (1973).
[11] R. Skomski, D. Leslie-Pelecky, R. D. Kirby, A. Kashyap, and D. J. Sellmyer, Scripta Mater. **48**, 857 (2003).
[12] R. D. Kirby, M. Yu, and D. J. Sellmyer, J. Appl. Phys. **87**, 5696-5699 (2000)

Mater. Res. Soc. Symp. Proc. Vol. 887 © 2006 Materials Research Society 0887-Q07-08

FAST AND SLOW MAGNETIZATION PROCESSES IN MAGNETIC RECORDING MEDIA

J. Zhou, R. Skomski, S. Michalski, R. D. Kirby, and D. J. Sellmyer
Department of Physics and Astronomy and Center for Materials Research and Analysis,
University of Nebraska, Lincoln, NE 68588

ABSTRACT

Information loss due to thermal activation is a major concern in ultrahigh-density magnetic recording media. The usually considered mechanism is thermally activated magnetization reversal over micromagnetic energy barriers. However, micromagnetic approaches ignore local anisotropy fluctuations, which translate into a time-dependent reduction of the remanent magnetization. The effect is negligibly small in macroscopic magnets but becomes important on a scale of a few nanometers.

INTRODUCTION

Magnetic recording has been a driving force in nanotechnology and electronics. The key advantage is the high storage density, corresponding to bit sizes much smaller than the wavelength of visible light. However, a fundamental bit-size limit is given by the thermal stability of the stored information [1]. In very small grains or particles, thermal activation leads to local magnetization reversal and to the decay of the stored information. In macroscopic magnets, there is a clear separation of time scales between fast atomic or intrinsic processes and slow extrinsic magnetization processes. Intrinsic properties can be treated by equilibrium statistical mechanics. For example, the local magnetic anisotropy $K_1(\mathbf{r})$ can be replaced by the time or ensemble average $<K_1(\mathbf{r})>$ [1, 2]. By contrast, extrinsic properties are related to hysteresis and often far from equilibrium [3, 4].

The question arises how magnetic systems behave on a length scale of a very few nanometers, where intrinsic phenomena become important. This applies, for example, to single-phase particles, core-shell structures, and exchange-coupled hard-soft structures, which have attracted renewed attention [5, 6] as magnetic-recording materials. This paper starts with a brief analysis of magnetization modes in the structures and then outlines how fluctuations of intrinsic properties affect the time-dependence magnetic properties.

NUCLEATION MODES IN COMPOSITE NANOPARTICLES

The simplest model of magnetization reversal, the Stoner-Wohlfarth model, describes uniformly magnetized small particles of volume V. In zero field, it yields the energy barrier $<K_1>V$ and a dynamics described by the Arrhenius law $\exp(-<K_1>V/k_BT)$. In two-phase particles (Fig. 1), magnetization inhomogenities are essential, because the soft phase switches earlier than the hard phase. Figures 1(b) shows the nucleation mode, that is, the magnetization deviation from $M = M_s\, e_z$ at the onset of magnetization reversal.

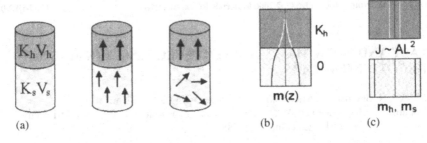

Fig. 1. Two-phase nanoparticle consisting of hard (h) and soft (s) phases: (a) structure, (b) magnetization modes, and (c) two-particle model. In (b), a magnetization tail reaches into the hard phase and reduces the energy barrier of the particle.

The model of Fig. 1(c) approximates the composite nanoparticle by two particles interacting via an effective exchange $J \sim 1/L^2$, where L is the particle size. The model has been used to discuss the coercivity of small permanent-magnet particles [7], but it can also be used to estimate the spin-wave dynamics of the system, in analogy to the well-known treatment of ferromagnetic resonance by Kittel [8, 9]. The solution of the problem amounts to the diagonalization of a 2×2 matrix whose nondiagonal matrix elements are proportional to J and where the demagnetization field of [8] is replaced by a more general anisotropy field. Figure 2 shows the predicted resonance frequencies. For large particles, $J \sim 1/L^2$ is negligible, and the system behaves like a superposition of hard and soft phases. On decreasing particle size, the modes hybridize and change their character. The lowest-lying mode can now be considered ferromagnetic, as in Fig. 1(b-c), whereas the excited mode is 'antiferromagnetic' with an oscillation 180° out of phase.

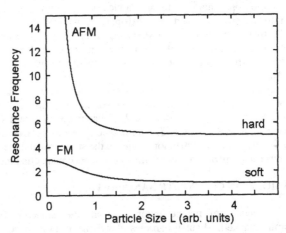

Fig. 2. Resonance modes in two-phase nanoparticles. The calculation is based on the model of Fig. 1(c).

ROLE OF FLUCTUATIONS

Micromagnetic approaches, such as Stoner-Wohlfarth model and the models of Fig. 1, start from thermally averaged intrinsic properties such as $<K_1(\mathbf{r})>$. This is a good approximation for macroscopic magnets but not necessarily for small nanoparticles. Figure 3 illustrates this point for a specific class of systems, namely rare-earth transition-metal intermetallics, for example $SmCo_5$ and $Sm_2Fe_{17}N_3$. In the intermetallics, the anisotropy is provided by the rare-earth sublattice, which consists of Sm^{3+} ions. It is well-known [10] that the finite-temperature average anisotropy $<K_1(T)> \sim (3<J_z^2> - J^2 - J)$ is caused by intramultiplet excitations ($J_z < J$) of the type shown in Fig. 3(b). On a micromagnetic level, this corresponds to the Arrhenius law $\tau = \tau_o \exp(-E_a(<K_1(T)>k_BT))$

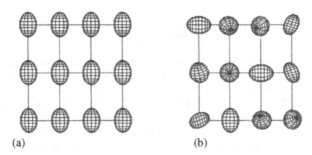

(a)   (b)

Fig. 3. Spin structure of rare-earth sublattices at (a) zero and (b) finite temperatures. The ellipsoids correspond to the 4f charge distribution of the Sm^{3+} ions, and the rare-earth moments are parallel to the axes of revolution.

However, nonzero fluctuations $<(K_1(\mathbf{r}) - <K_1>)^2>$ facilitate the magnetization reversal by reducing the energy barriers, rather than driving the magnetization over the energy barrier. Figure 4 illustrates that the size of the maximum or 'giant' fluctuations increases with waiting time. In macroscopic systems, these fluctuations can be ignored, but in very small particles, they affect the long-time behavior of the magnet and add to the magnetization decay predicted from $<K_1>$. The effective anisotropy $K_{eff} = <K_1> - \delta K$ obeys

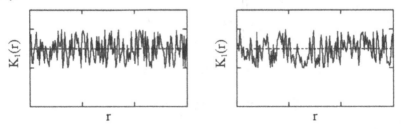

Fig. 4. Anisotropy fluctuations. Giant fluctuations, such as the negative fluctuation near the center of the right figure, require long waiting times.

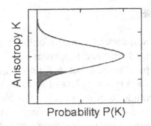

Fig. 5. Gaussian anisotropy distribution, averaged over a volume ξ^{3d}. The gray area is inversely proportional to the waiting time.

$$K_{\text{eff}}(T, t) = <K_1> - \delta K_o \sqrt{2 \ln(t/\tau^*)} \qquad (1)$$

where δK_o and τ^* depend on temperature and one the volume over ξ^{3d} over which the anisotropy is averaged. As it will be explained elsewhere, Eq. (1) involves a complicated random-anisotropy and time averaging over all spin configurations. In a simple interpretation, Fig. 5, the logarithmic time dependence of K_{eff} reflects the low-lying parts of the Gaussian anisotropy distribution, which is a consequence of the mechanism of Fig. 3(b).

To quantify the effect in an approximate manner, we have performed a numerical simulation using the OOMMF code by NIST (http://math.nist.gov/oommf). Thermal excitations enter the calculation in form of a random anisotropy whose magnitude depends on the waiting time. In other words, we temporarily 'freeze' the dynamic spin configuration to check whether the thermal disorder is sufficient to realize magnetization reversal in a given reverse field. The investigated cylindrical single-phase particle has a diameter of 8 nm and a height of 10 nm. The used materials parameters are $<K_1> = 5$ MJ/m³, $A = 10$ pJ/m, and $M_s = 1100$ MA/m. In the simulations $\delta K = 1 - 3$ MJ/m³. Figure 6 shows the hysteresis of the particle for different waiting times (sweep rates).

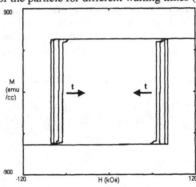

Fig. 6. Time-dependence of the hysteresis loop in the 'frozen' or 'snapshot' approximation.

DISCUSSION AND CONCLUSIONS

It is instructive to compare the present local-anisotropy picture to the Stoner-Wohlfarth reversal in small particles. In the Stoner-Wohlfarth picture, Fig. 7(a), the spins remain parallel during magnetization reversal, but spin disorder (b) opens another channel for reversal. Usually, the random contribution is small, but in particles smaller than 10 nm, it may become nonnegligible. For example, in thins wires of cross section πR^2, the activation volume scales as R^2 and approaches zero for very small R. In this limit, local randomness becomes the main consideration.

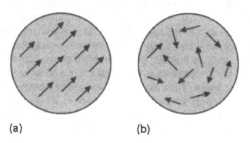

(a)   (b)

Fig. 7. Comparison of Stoner-Wohlfarth model (a) and atomically randomized spins (b).

Naturally, our quasistatic or 'frozen' Gaussian approximation is relatively crude. It can be shown that Eq. (1) requires some time averaging to ensure that the giant anisotropy fluctuation exists for about 1 ns, so that thermal reversal can occur. In Eq. (1), this averaging appears in for of the time constant τ^*, which is larger than the time scale of atomic anisotropy fluctuations and reduces δK. This procedure ignores, for example, fast

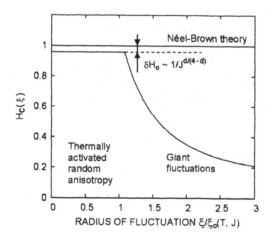

Fig. 8. Coercivity as a function of the size of bulk anisotropy fluctuations.

magnetization processes that have the character of spin precessions. The corresponding fast magnetization dynamics in nanostructures and multilayers is currently under experimental and theoretical investigation, but its discussion goes beyond the scope of this paper.

To gauge the effect of the anisotropy fluctuations, it is also necessary to consider their size ξ. Atomic-scale spin disorder occurs frequently but has very little effect on the anisotropy. To have a significant effect on the magnetization reversal, the fluctuation must comprise many atoms. In the simulations leading to Fig. 6, this effect is automatically included. Analytically, it is necessary to nucleation field in a random potential, as outlined in [3]. Figure 8 shows the corresponding result.

In conclusion, we have investigated how local anisotropy fluctuations affect the magnetization dynamics of small particles. Our preliminary numerical and analytic calculations quantify the effect by using the 'frozen' or 'snapshot' approximation, which maps the finite-temperature problem onto a problem onto a static but disordered system. The local fluctuations are most pronounced in small grains or particles and negatively affect the thermal stability. They amount to fluctuating energy barriers and reduce the remanent magnetization beyond the Arrhenius-Néel-Brown predictions. The effect, which we have treated in a static approximation, adds to the dynamics based on micromagnetic free energies, such as Arrhenius-Néel-Brown activation and ferromagnetic resonance.

ACKNOWLEDGEMENT

This research is supported by the W. M. Keck Foundation, NSF-MRSEC, INSIC, and CMRA.

REFERENCES

[1] D. Weller and A. Moser, IEEE Trans. Magn. **35**, 4423-4439 (1999).
[2] D. J. Sellmyer and R. Skomski (Eds.), "Advanced Magnetic Nanostructures", Springer, Berlin 2005.
[3] R. Skomski, J. Phys.: Condens. Matter **15**, R841 (2003).
[4] D. Givord and M. F. Rossignol, in: "Rare-Earth Iron Permanent Magnets", Ed.: J. M. D. Coey, University Press, Oxford 1996, p. 218-285.
[5] R. H. Victora and X. Shen, IEEE Trans. Magn. **41**, 537 (2005).
[6] D. Suess, T. Schrefl, S. Fähler, M. Kirschner, G. Hrkac, F. Dorfbauer, and J. Fidler, Appl. Phys. Lett. **87**, 012504 (2005).
[7] J. M. D. Coey and R. Skomski, Physica Scripta **T49**, 315 (1993).
[8] C. Kittel, "Introduction to Solid-State Physics", Wiley, New York 1986.
[9] M. I. Chipara, R. Skomski, and D. J. Sellmyer, J. Magn. Magn. Mater. **249**, 246 (2002).
[10] R. Skomski and J. M. D. Coey, "Permanent Magnetism", Institute of Physics, Bristol 1999.

Degradation Processes Induced by
Particles and Ionizing Radiation

Mater. Res. Soc. Symp. Proc. Vol. 887 © 2006 Materials Research Society 0887-Q10-07

Radiation Degradation of Nanomaterials

R.P. Raffaelle[1], Cory D. Cress,[1] David M. Wilt,[2] and S.G. Bailey[2]

[1] Rochester Institute of Technology, Rochester, NY 14623
[2] NASA Glenn Research Center, Cleveland, OH 44135

ABSTRACT

We have been developing a variety of nanomaterials for their use in power devices. An example of this is our use of both single wall carbon nanotubes and several varieties of semiconducting quantum dots (e.g., $CuInS_2$, CdSe, InAs) for use in space solar cells. The ability of these materials to withstand the rigors of the space radiation environment will be essential for this intended application. In addition, we have also been developing both nanostructured III-V devices and radioluminescent quantum dots for use in radioisotope batteries. In this application these nanomaterials are subjected to an extremely high radiation level. Their degradation rate will be the key to determining the ultimate lifetime of these power supplies, which in principle can have an energy density that is orders of magnitude higher than any conventional battery chemistry. The nanomaterials included in this study were subjected to alpha particles fluences and the degradation of various properties was monitored using different analytical techniques. Specifically, the radioluminescence of the quantum dots intended for use in the radioisotope batteries was monitored as a function of fluence. In the case of the III-V quantum dots, their photoluminescent degradation as a function of fluence was measured in comparison to the bulk substrate on which the quantum dots were grown. Finally for the carbon nanotubes, relative intensities of the Raman peaks associated with their inherent vibrational modes were used to monitor the effects of the alpha radiation damage. Results on the radiation tolerance of these nanomaterials and its implication with regard to their ultimately utility in power devices are presented.

INTRODUCTION

The use of nanomaterials in power generation and storage has quickly become one of the most exiting areas of science and technology today [1-3]. Many of these applications involve either radiation exposure directly or operation in high radiation environments (i.e., solar or radioisotope exposure [4], space environments [5], etc.). The behavior of nanomaterials, such as single wall carbon nanotubes and semiconducting quantum dots, in these environments is a very important consideration. Radiation tolerance may be as important as many of their other unique and remarkable properties for a host of applications.

Carbon nanotubes have been found to possess a wide variety of extremely remarkable properties, most notably high electrical and thermal conductivity, mechanical strength, and catalytic surface area [6]. These properties imbue carbon nanotubes with tremendous potential for a variety of power generation and storage devices including: Lithium-ion (Li^+) batteries, polymeric solar cells, proton exchange membrane (PEM) fuel cells, and thermionic power devices [1]. In addition, carbon nanotubes show size-dependent optoelectronic properties based upon quantum confinement which could prove extremely useful in terms of solar energy applications. Similarly, the exploitation of quantum confinement effects in semiconductors has generated an entire new way of thinking about semiconductor device design.

In the past, when designing semiconductor optoelectronic devices, we were somewhat restricted by the various properties of bulk semiconductors. Compositional changes were the primary means of "tuning" the materials to meet certain need. Recently, scientists have started to utilize material size as a method of controlling the optoelectronic properties of materials. When the size of a material becomes lower than the Bohr exciton radius of an electron-hole pair in the material (i.e., the nanoscale), it will start to exhibit size-dependent optoelectronic properties. When the electrons in a material are quantum confined in all three dimensions, we refer to that material as a quantum dot. These nanocrystallites behave essentially as a 3-dimensional potential well for electrons (i.e., the quantum mechanical "particle in a box").

Theoretical results of Luque and Marti have shown that a photovoltaic device with an intermediate band of states, such as resulting from the introduction of quantum dots, can exceed the Shockley and Queisser model efficiency of not only a single junction but also a tandem cell device [7]. By introducing an ordered array of uniform quantum dots within the intrinsic region of a p-i-n solar cell an ideal efficiency of 63% is predicted. Quantum dots have already been used successfully to improve the performance of devices such as lasers, light emitting diodes, and photodetectors [8]. In addition to optoelectronic tuneability, it has also been shown that quantum dots could have some benefits in terms of radiation tolerance [9]. This is especially important in the development of solar cells for space.

There are several approaches to mitigating the effects of high radiation on solar arrays in space. The simplest is to employ thick cover glass (assuming that a commercial source could be developed). Thick cover glasses will cause a significant decrease in the specific power of the array and increase the mass of the system. However, this can be reduced if one adopts a concentrator design, assuming of course that the additional elements associated with the concentrator can withstand the high radiation environment as well. A different approach is to develop cells that are more radiation-resistant. The possibility of utilizing quantum dots to improve the radiation tolerance in addition to cell efficiency is extremely attractive for space power application. Carriers confined to a quantum dot do not interact with radiation induced defects in the same manner that free carriers in bulk semiconductor do [9]. Therefore, the rate of degradation that is normally seen after exposure to a fluence of high-energy particles, in many devices, can be reduced through the incorporation of quantum dots. In addition to space solar cells, another technology for which the radiation tolerance of quantum dots and other nanostructured materials may hold great benefit are radioisotope batteries.

Radioisotope batteries convert the kinetic energy from the alpha or beta-particles, emitted from the nuclei of unstable atoms, into usable electricity. The energy density that is available in a radioactive source far exceeds any chemical or gas battery. The technological barrier to their development has been the radiation degradation of the materials used to convert emitted particles to electricity. A simple example of a radioisotope battery can be constructed by placing a radioisotope source in contact with an ordinary semiconductor p-n junction or solar cell. The high energy particles create electron-hole pairs in an analogous way to light from the sun. However, unlike a photon from the sun, one alpha particle, for example, can create thousand of carriers. This is apparent when considering a single alpha particle from Am^{241} has 5.5 MeV of energy and the bangap of Si, for example, is merely 1.1 eV. Unfortunately, an ordinary solar cell p-n junction is a diffusion dominated device. Its performance is inextricably linked to the minority carrier lifetimes in the semiconductor. The minority carrier lifetime is what is most susceptible to radiation degradation. A new approach [4] which could be enabled by the radiation tolerance of quantum dots in which they are used as an intermediate absorber material (see Figure 1).

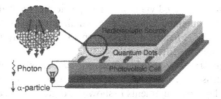

Figure 1. Schematic of a radioisotope battery with an intermediate quantum dot absorber.

By placing quantum dots between a radioisotope source and a solar cell, which is tuned to the radioluminescence of the quantum dots, one can obviate the problem of radiation damage in the solar cell. Of course, the overall efficiency of the device will be the product of the radioluminescent efficiency of the quantum dots and photovoltaic conversion efficiency of the cell. However, with the tremendous energy density present in the source it is reasonable to give up some efficiency for longevity in such a device. The issue that ultimately determines the efficacy of such an approach is the radiation tolerance of the quantum dots.

EXPERIMENTAL DETAILS

Doped zinc sulphide (ZnS), yttria (Y_2O_3), and yttrium aluminum garnet (YAG) nanocrystals were prepared for this study. A pulsed laser deposition technique was utilized in the synthesis of our ZnS:Ag,Cl nanophosphor material. This process began by preparing a 4:1 mixture of ZnS:Ag (Saint Gobain) to NaCl (Sigma Aldrich). The mixture was homogenized via mortal-and-pestle and mechanical stirring. This material was then pressed at 30,000 psi forming a 2" target with a mass of 2.5 g. The target was placed at the focal point of an alexandrite laser (λ =755 nm) within a sealed quartz tube. After evacuating and purging with Ar 3x, a horizontal tube furnace was used to heat the reaction vessel to 700° C. Laser deposition was then performed by rastering the laser emission over the surface of the target. A continuous flow of Ar (100 sccm) was used to transport the material down the length of the furnace, where it finally condensed on the walls of the quartz tube. The reaction vessel was kept at atmospheric pressure. NaCl was used to help bind the target (note: the Cl acts as a co-activator increasing the radioluminescence and provides a slight red-shift in the emission wavelength).

A simple ethylenediaminetetraacetic acid (EDTA) assisted urea precipitation technique was used to synthesize Y_2O_3:Re (Re=Eu^{3+}, Er^{3+}, Tb^{3+}) nanophosphors with precise size control. This process begins by dissolving 10 g of Y_2O_3 and a stochiometric amount of Re_2O_3 in 100 mL of 3M nitric acid, resulting in an aqueous solution of $Y_xRe_{1-x}(NO_3)_3$ with a pH of ~2. 12 mL of this stock solution is then diluted 25 fold in H_2O and 15 g of urea is added. Depending on the desired crystal size, 0-1.0 g of EDTA is added to the solution, after which the solution is refluxed for 2 hours. The solution is then filtered using a 0.02 μm anodisc alumina filter, rinsed

3x with H_2O, and dried in vacuo at 80° C. After drying, a fine powder results which is crushed with a mortal-and-pestle and fired at 1150° C in air.

A sintering process was used in the preparation of rare-earth doped YAG nanophosphors. The process began by mixing a stochiometric amount (Y:Re) of Re_2O_3 nanopowder (99.5%, Sigma Aldrich) to YAG nanopowder (99% Sigma Aldrich). To this mixture, BaF (99.999%, Sigma Aldrich) was added as a flux material in a 1:3 (BaF to YAG:Re) ratio. The mixture was homogenized via mortal-and-pestle and mechanical stirring followed by sintering in air at 1200° C contained in loosely sealed alumina boats, for 6 hours. The sintered material was washed in a 1.5 M nitric acid solution to remove any residual BaF and filtered using a 0.02 μm anodisc alumina filter.

The resulting materials were characterized by scanning electron microscopy and photoluminescence spectroscopy. The radioluminescence of the materials was also measured and the setup for taking these measurements is shown in Figure 2. A sample of the nanocrystals was prepared by dispersing the material in methanol and dispensing on a quartz slide. The light emitted from the phosphor, when excited by the radioisotope source, is then measured by a radiometer attached to the fiber optical cable.

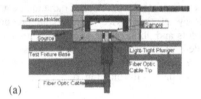
(a)

(b)

Figure 2. a) Schematic of the source and sample holder used for the radioluminescence measurements and b) a photograph of the Isotope Products Laboratory source used in the measurement.

The InAs III-V quantum dots used in this study were grown in a horizontal, reduced pressure, organomettallic vapor phase epitaxy (OMVPE) reactor. Metallorganics, trimethyl gallium (TMGa) and trimethyl indium (TMIn), were used as precursor materials, along with hydrides phosphine (PH_3), and 1% AsH_3 in hydrogen. Disilane (10^2 ppm Si_2H_6) and dimethyl zinc (10^3 ppm DMZn) diluted with hydrogen were used as sources of n- and p-dopants, respectively. The typical growth temperatures for the GaAs were 675ºC. All growth runs were performed at a pressure of 600 Torr. The technique used to produce the quantum dots is called Stranski-Krastanow growth. This type of growth occurs for crystals of dissimilar lattice parameters, but low interfacial energy (i.e., InAs on GaAs). After an initial layer-by-layer growth, islands form spontaneously, leaving a thin "wetting layer" underneath. In contrast to Volmer-Weber systems, where decreasing the surface energy reduces islanding by forcing deposited material to wet the substrate, in Stranski-Krastanow systems the substrate is already "wetted", so decreasing the surface energy increases islanding since bulk strain energy is lowered. This method has been successfully used to grow a wide variety of III-V quantum dots for a number of different electronic applications [10-11]. The InAs quantum dot morphology was characterized by atomic force microscopy using a Digital Instruments Nanoscope-E scanning probe microscope. The degradation in the photoluminescence of the quantum dots as a function of fluence was measured using a Jobin Yvon LABRAM HR 800 photoluminescence microscope.

Synthesis of the single wall carbon nanotubes (SWNTs) used in this study was performed using an Alexandrite laser vaporization process (see Figure 3), previously described in detail [12]. The raw soot was purified using conventional nitric acid and thermal oxidation steps, to achieve SWNT mass fractions of >95% w/w in the overall sample [12]. The SWNTs were characterized via thermogravimetric analysis and optical spectroscopy to assess their residual metal content and purity. The SWNTs were imaged using a Hitachi S-900 field emission scanning electron microscope. The degradation of the SWNTs with respect to alpha particle fluence was monitored using Jobin Yvon LABRAM HR 800 Raman microscope.

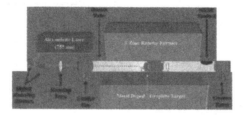

Figure 3. Schematic of single wall carbon nanotube laser vaporization reactor.

DISCUSSION

Figure 4 shows the variation in particle size of the Er doped Y_2O_3 nanoparticles as the amount of EDTA is varied in the synthesis. Smaller crystals were obtained by increasing the EDTA concentration during synthesis. Un-reacted precursor material or second phase was present in the samples generated without the addition of the EDTA. The smallest Y_2O_3 nanocrystals, along with the doped YAG nanocrystals were excited by the 488 nm line from an Argon Ion laser and their subsequent photoluminescence was measured (see Figure 5). They all showed spectrally rich photoluminescence in the 550 to 650 nm range. Assuming that their radioluminescence spectrum is similar it would make them all suitable candidates for a radioisotope battery which utilized an InGaP photovoltaic diode. This is due to their emission energies being above the 1.8 eV bandgap of InGaP grown lattice-matched to GaAs.

Figure 4. SEM micrographs of Y_2O_3:Er^{3+} nanocrystals synthesized by the UREA precipitation technique with a) No EDTA and (b) and (c) with increasing amounts of EDTA.

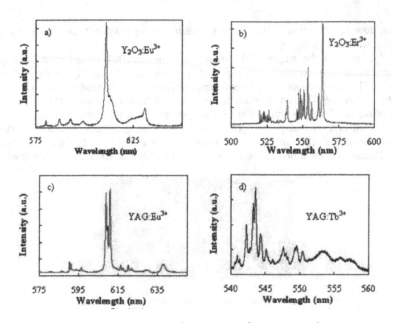

Figure 5. Photoluminescence of a)Y2O3:Eu^{3+}; b)Y2O3:Er^{3+}; c) YAG:Eu^{3+}; and d) YAG:Tb^{3+}. The nanophosphors were excited by a 488 nm argon laser.

To determine the appropriate quantum dot film thickness for a radioluminescence measurement (note: thickness was not critical for the PL as it was done in a reflectance mode) we performed Transport of Ions in Matter (TRIM) calculations (see Figure 6). This technique uses a Monte Carlo approach to determine the path of the incoming alpha particles in various materials. The results show that for all phosphors synthesized, a film of approximately 12 μm is sufficient to absorb the majority of the incoming particles and still limit the self-absorption of the resulting photons which are created in the film.

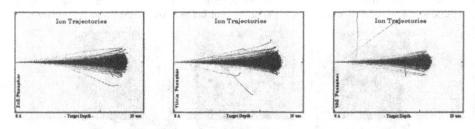

Figure 6. TRIM Ion trajectory Monte Carlo simulation results for ZnS phosphors, Y2O3 phosphors, and YAG phosphors.

The radioluminescence (RL) of the same phosphors which were used in the photoluminescence (PL) measurements were measured using the setup in Figure 2. The results for the doped YAG nanocrystals and the doped ZnS are shown in Figure 7. The peak intensities of the emission spectra where shifted in the RL in comparison to the PL. This is not too surprising considering the disparity in the excitation energies between the Argon Ion laser produced photons and the 4.2 MeV alpha particles from the Po^{210}. The alpha particles are capable of populating the higher energy levels associated with the rare earth dopants thus altering the energy transitions of the electrons and their associated emission spectrum. Peak emission from the doped ZnS is blue shifted in respect to the other phosphors making it less attractive for a radioisotope battery application using an InGaP converter. The Tb doped YAG is an excellent candidate as is shown in the Figure 8 overlay of the spectral response of a 1.8 eV InGaP photovoltaic diode with the YAG:Er^{3+} radioluminescence spectrum.

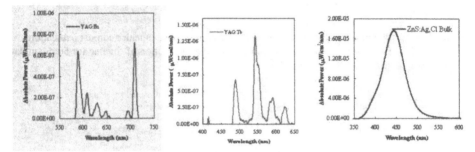

Figure 7. Radioluminescence Spectra of YAG:Eu3+, YAG:Tb3+, and ZnS:Ag,Cl.

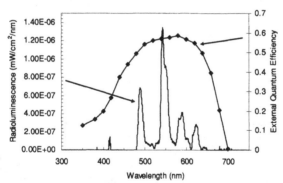

Figure 8. Overlay of the spectral response of a 1.8 eV InGaP photovoltaic diode and the radioluminescence spectrum of YAG:Tb^{3+}.

The degradation of the PL intensity of the phosphors was monitored as a function of fluence. Figure 9a shows that up to a fluence of 10^{11} alphas there was very little observed PL reduction for the Er doped Y_2O_3. Examining the radioluminescence degradation for the same phosphor as a function of size showed that the materials with an average particle size of 50 nm were significantly more radiation tolerant than their 100 nm counterparts (see Figure 9b). Figure 10 shows the radioluminescence degradation for the nanoscale versions of the other phosphors with the YAG:Tb showing a 1.3% degradation over of fluence of 10^{12} 4.2 MeV alphas

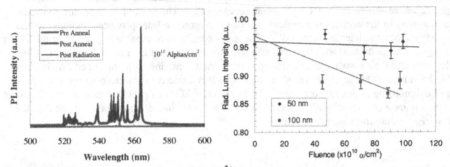

Figure 9. a) Photoluminesce spectra of Y_2O_3:Er^{3+} pre anneal, post anneal, and post irradiation with 10^{11} 4.2 MeV alphas/cm^2 and b) radioluminescence as a function of fluence for 50 nm and 100 nm average particle size Er^{3+} doped Y_2O_3.

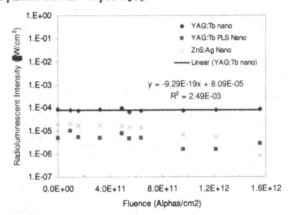

Figure 10. Radioluminescence versus fluence for two different YAG:Tb and a ZnS:Ag phosphor. The line is a least-square fit. (PLS -pulsed laser synthesized).

Figure 11a shows an atomic force micrograph of InAs quantum dots produced by OMVPE. These quantum dots were uniformly deposited on the GaAs surface and had an average particle size (height) which ranged from 5 to 10 nm. The photoluminescence of the InAs quantum dots was measured as a function of increasing fluence (see Figure 11b). Both the PL peak attributed to the QDs and that of the bulk GaAs decreased with fluence. However, using a 4 - Gaussian fit to the data enabled the examination of the degradation of the individual peaks as a function of fluence (see Figure 11 c). The peak intensities as a function of fluence are given in Figure 11d. The curves show the relative decrease in the photoluminescence intensity of the QD ground state in comparison to the bulk GaAs.

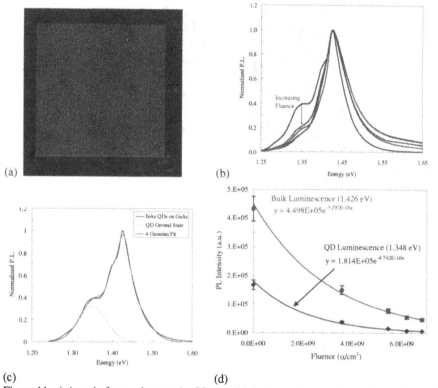

(a)   (b)   (c)   (d)

Figure 11. a) Atomic force micrograph of Stranski-Krastonov InAs quantum dots on GaAs; b) normalized photoluminescence overlays for differing fluences; c) a 4-gaussion fit to a single normalized PL showing the Gaussian peak that would correspond to the QD ground state; and d) the PL intensity as a function of fluence for the ground state of the QDs and of the bulk GaAs.

Figure 12 shows field emission scanning electron micrographs of the SWNTs produced by laser vaporization (a) before and (b) after purification by acid refluxing and thermal treatment. The images demonstrate the effectiveness of the purification evidenced by the removal of the catalyst metal atoms and non-tubular carbonaceous materials. Figure 13a shows the Raman spectrum of the purified SWNTs with overlays for increasing fluence. A drop is seen in the intensity of the various peaks. Comparison of the peak ratios of the D to G^+ and the G^+ to G' band shows that initially there is a slight increase in the D to G^+ ratio as would be expected. Also, that the ratio does start to decrease eventually after a fluence of approximately 10^{13} 4.2 MeV alphas. However, the most striking change is in the ratio of the G^+ to G' bands. It increases linearly with fluence over the entire range measured. This would indicate a larger decrease in the Raman peak intensity associated with G' band in comparison to the other bands. This would indicate that the radiation defects being created are extremely detrimental to the two-phonon resonances which bring rise to the G' band.

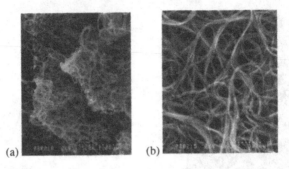

Figure 12. Field emission scanning electron micrographs of a) laser produced SWNT raw material and b) after purification.

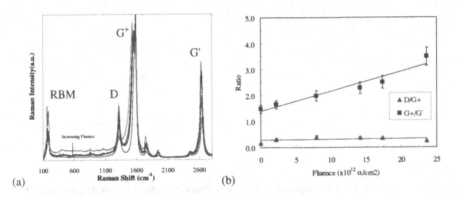

Figure 13. a) Raman spectrum of purified laser SWNTs overlays as a function of fluence and b) the Raman peak ratios as a function of fluence.

CONCLUSIONS

Nanostructured approaches can be used to improve the radiation tolerance of the materials used in radioisotope battery applications, and thus may provide the means to take advantage of radioisotope energy densities for long-lived microbatteries. The photoluminescence of SK grown InAs quantum dots on GaAs decreases exponentially with increase alpha fluence. The rate of decline is less than that of the bulk GaAs as anticipated. Single wall carbon nanotubes do degrade under high energy alpha flux. However, a smaller than anticipated increase in the D band is observed, whereas a large decrease in the G' band in relation to the other bands is measured.

ACKNOWLEDGEMENTS

We would like to acknowledge the support of the NASA Human and Robotics Technology (H&RT) Program Grant #NNC056A60G, the DARPA RadioIsotope Micropower Sources (RIMS) program Grant #NBCH1040010, BP Solar, Inc., Alpha V, Inc., and the National Science Foundation Grant #ECS-0233776.

REFERENCES

[1]. R. P. Raffaelle, B.J. Landi, J.D. Harris, S.G. Bailey, A.F. Hepp, *Mat. Sci. Eng. B*, *B116*, 233 (2005).

[2]. B.J. Landi, S.L. Castro, H.J. Ruf, C.M. Evans, S.G. Bailey, and R.P. Raffaelle, *Sol. Energy Mater. Sol. Cells*. **87**, 733-746 (2005).

[3] S. Castro, S. Bailey, R.P. Raffaelle, K. Banger, and A. Hepp, *J. Phys. Chem. B*, **108**, 12429-12435, (2004).

[4]. S.G. Bailey, D.M. Wilt, S.L. Castro, C. Cress and R.P. Raffaelle, *IEEE Photovoltaics Specialists Conference*, Orlando, FL (2005).

[5]. R.P. Raffaelle, S.L. Castro, A.F. Hepp, and S.G. Bailey, *Prog. in Photovolt.*, **10**, 1 (2002).

[6]. H. Dai, *Surf. Sci.* **500**, 218-241 (2002).

[7]. A. Luque and A. Marti, *Phys. Rev. Lett.*, 78, 26 (1997).

[8] R. Leon, *Appl.Phys. Lett.* 76, 15 (2000)

[9] S. Marcinkevicius, *Phys. Rev. B.* **66**, 235214 (2002).

[10] S. Kamprchum, S. Kiravittaya, R. Songmuang, S. Thainoi, S. Kanjanchuchai, M. Sawadsaringkam, and S. Panyakeow, *IEEE Photovoltaics Specialists Conference*, New Orleans, LA, May 20-24 (2002).

[11] P.M. Petroff and S.P. DenBaars, Superlattice Microstructures, 15, 15 (1994).

[12] B.J. Landi, R.P. Raffaelle, S.L. Castro, S.G. Bailey, *Prog. in Photovolt* 13, 1-8 (2005).

Mater. Res. Soc. Symp. Proc. Vol. 887 © 2006 Materials Research Society   0887-Q10-08

Degradation of Nano-crystalline ITO Films Due to Exposure to Hyperthermal Atomic Oxygen

Long Li[1], Ross Harder[2], Fengting Xu[1], Ian K. Robinson[2] and Judith C. Yang[1]

1. Materials Science and Engineering Department, University of Pittsburgh, Pittsburgh, PA 15261.
2. Department of Physics, University of Illinois at Urbana-Champaign, Urbana, IL 61801, USA.

ABSTRACT

Indium tin oxide (ITO) films coated on float glass slides were exposed to 5 eV hyperthermal atomic oxygen at room temperature with increasing fluences: 2×10^{19}, 6×10^{19} and 2×10^{20} O-atoms/cm^2. We characterized the structure of the ITO films after room temperature atomic oxygen exposure with scanning electron microscope (SEM) and atomic force microscope (AFM), synchrotron X-ray diffraction (XRD), and cross-sectional transmission electron microscope (X-TEM). The unexposed ITO films were found to possess a nano-crystalline surface, and clean and abrupt ITO/SiO$_2$ interfaces without interfacial phase. Surface roughness of the exposed ITO films increased with the increasing AO influences. The interface- sensitive peaks in XRD measurements with grazing incidence revealed that the crystallinity of the ITO was modified near the interface. Cross-sectional TEM confirmed that many ITO particles with diameters ranging from 2-10 nm formed in the SiO$_2$ substrate near the interface after AO exposure. These findings suggest that O atoms can travel through the ITO films, where the boundaries of columnar-grown grains may supply the pathway.

INTRODUCTION

Spacecraft encounter a harsh environment in the low-Earth orbit (LEO), ranging from approximately180 to 650 km above the Earth[1-3]. The primary hazardous species in the LEO are vacuum ultraviolet (VUV) radiation, X-rays, atomic oxygen which is formed by the photodissociation of molecular oxygen, and electrons and other charged species[1,2]. The high relative velocity between space vehicles and the ambient atomic oxygen (AO) leads to hyperthermal collisions of O atoms with spacecraft materials, with 5 eV kinetic energy[4,5]. Indium-tin-oxide (ITO), a transparent conductive compound oxide, is typically employed as a coating material to protect Kapton blankets, used in solar arrays in LEO, from AO-erosion and bleed off static charges from the insulating polymer surface[6]. However the ITO coatings did degrade when exposed on the ram side of a space vehicle in the LEO environment[7,8]. Ground simulations of AO reaction with ITO were conducted, however, Synowicki *et.al* and Woollam *et.al.* reported non-effective of their thermal asher AO source due to the low kinetic energy of their source (0.5 eV)[7,8]. We used our hyperthermal AO source to expose ITO films coated on float glass to simulate the LEO on the ground. ITO films were exposed to 5 eV AO at room temperature and structure characterizations were carried out using SEM, AFM and synchrotron XRD, and cross-sectional TEM and high-resolution TEM (HRTEM), high angle annular dark field (HAADF) and energy dispersive spectroscopy (EDS).

EXPERIMENTAL

ITO films ($90w\%In_2O_3$: $10w\%$ SnO_2) were sputter-coated on SiO_2-passivated float glass-slides in the vacuum (10^{-6} Torr level). The ITO films were cleaned with isopropyl alcohol and then rinsed with deionized water several times. After air-drying in a clean room, the ITO samples were stored in a vacuum desiccator.

The 5 eV atomic oxygen was created by the laser detonation of oxygen gas[9]. The AO source utilizes a high power pulsed CO_2 laser to ignite a plasma breakdown of the O_2 gas. The rapidly expanding plasma dissociates the molecular oxygen into a neutral atomic oxygen beam. The fraction of atomic oxygen in the beam exceeded 80%. An internal bakeout halogen IR lamp was mounted inside the chamber to shorten pumping-down time to reach ultra-high vacuum of 10^{-9} torr. ITO samples were exposed to 5 eV AO beam at room temperature with increasing fluencs of 2×10^{19}, 6×10^{19} and 2×10^{20} O-atoms/cm^2, respectively. Witness samples of Kapton H polyimide were employed to calculate the total AO fluence [5]. After exposure, the samples were cut into several small pieces for SEM and AFM, synchrotron XRD, and cross-sectional TEM studies.

Two ITO pieces of 1.6×10 mm (1.2mm in thickness) were cut off from the exposed sample, and glued face-to-face with Gatan G1 glue and cured at 75 °C for 1 hour to make a "sandwich". The sandwich was sliced to 300-micron rectangle blocks with a diamond wire saw. These rectangle pieces were ground and polished on both sides to 100 microns with SiC sandpaper, dimpled on both sides to 25 microns with a Gatan 656 dimpler, and then Ar-ion-milled with a Gatan 691 PIPS. The voltage used for the ion mill is 4.5 kV and the milling angle was around 4°.

The surfaces of the ITO samples were examined with Philips XL-30 FEG (field emission gun) SEM and Digital Instruments DI3100 AFM in tapping mode. The Synchrotron XRD measurements were conducted with grazing incidence using the beamline 34_ID-C at the Advanced Photon Source (APS), Argonne National Lab. The high intensity of the APS permits direct imaging the powder diffraction of the ITO films with CCD (charge coupled device). The cross-sectional observation through the surface to the interface of ITO films unexposed and exposed was carried out with JEOL 2000FX TEM and 2010 FEG (S)TEM.

RESULTS AND DISCUSSION

Figure 1 is images of the ITO samples before AO exposure. Figure 1a is a SEM image which shows the ITO grains in bundles on the surface. The two typical kinds of nano-sized grains, one with round shape in diameters ranging 20-50 nm and the other one with needle-like shape of 20-50 in width by ~100 nm in length. Figure 1b is AFM image showing the surface of original ITO film with an RMS roughness of 5.3 nm. Figure 1c is a cross-sectional TEM image which showed the columnar structure of the ITO film in a 200 nm thickness. The inset selective area electron diffraction pattern (SAD) showed polycrystalline ITO film. A layer of amorphous SiO_2 shown in figure 1c, is approximately ~40 nm in thickness. The interface of ITO/SiO_2 is abrupt without interfacial phase. Figure 1d is an HRTEM image of a typical grain boundary. The broad transition area at the boundary indicates non-contacting boundary of the finger like grains. The elemental composition by energy dispersive spectroscopic (EDS) in the TEM demonstrated that the film is composed of O, In and Sn only.

Figure 1. Structural characterization for ITO films before AO exposure. Figure 1a and 1b are SEM and AFM images of ITO film surface; figure 1c is cross-sectional TEM image and SAD pattern; 1d is an HRTEM image, showing a grain boundary between two columnar grains.

Figure 2 is a set of SEM images of the ITO surfaces after AO exposure. Figure 2 a, b, c and d were from the surface with fluences of 0 (unexposed), 2×10^{19}, 6×10^{19} and 2×10^{20} atoms/cm^2, respectively. The dark contrast shows the damaged surface after AO exposure. The fractional coverage of the damaged area for the surfaces of b, c and d was approximately 18%, 25% and 15%, respectively. The coverage did not depend on the AO fluences in AO exposures, and no direct evidence is provided that it is in fact due to original imperfections of the ITO surface. We also measured the surface roughness (RMS) on the ITO surfaces (table 1) with AFM. The surface roughness increases with the increasing fluences from 2×10^{19} to 2×10^{20} atom/cm^2. When comparing to the exposed surface to the unexposed surface, the surface roughness reduced from 5.34 nm to 4.7 nm. This could be due to the removal of the surface contamination [10]. The reactions of ITO films with AO not only changed the surface roughness, but also modified the region under the surface as observed with SEM by surface topographic technique.

Figure 3 is intensity profiles of the power ring at 2.5Å^{-1}, recorded with a CCD of synchrotron XRD, as a function of exit angle collected from ITO films with creasing AO fluences. The interface-sensitive peak of XRD at grazing incident angle of 0.7° contributed from grains located deeper in the film. The red curve shows the integrated intensity profile of unexposed film. It's relatively flat character shows that the film is fairly homogeneous. The green and blue profiles

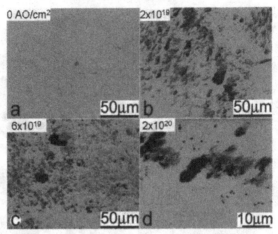

Figure 2. SEM images of a set of ITO surface exposed with increasing AO fluences.

Table 1. Surface roughnesses of ITO films exposed to AO

Figure 2	a	b	c	d
Fluence (atom/cm^2)	0	2×10^{19}	6×10^{19}	2×10^{20}
Roughness, RMS (nm)	5.34±0.07	4.7±0.04	5.1±0.08	6.1±0.12

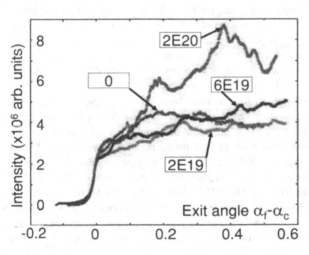

Figure 3. XRD integrated intensity profiles as a function of exit angle for ITO films with increasing AO fluences.

show evidence of greater crystalline structure within the film with AO fluence of 2×10^{19} and 6×10^{19} atom/cm^2. The sharp peaks in the profile of 2×10^{20} atom/cm^2 AO fluence, shown in purple, illustrate a significant modification of the film by AO at the interface. The greatest peaks occur at the higher exit angles supplied evidence that the larger crystallites are localized deep within the film, near the interface. To prove it, we chose the ITO sample with the largest fluence of 2×10^{20} atom/cm^2 for cross-sectional TEM. Many pieces of cross-sectional samples were made for better statistics. Figure 4(a) is a bright field TEM image of the interface of ITO/SiO$_2$. Many particles appeared in the SiO$_2$ substrate near interface, as compared to the clean interface of the unexposed ITO film (figure 1c). The particle sizes ranges from 2 to 10 nm. On the HRTEM image in figure 4(b), these particles are well-crystallized, whose lattice parameters fit to that of the ITO JCPDS file. The particles were characterized with energy dispersive spectroscopy of x-ray (EDS), as shown in figure 4(d) with the reference HAADF image (figure 4(c)). Spectra 1 to 6 were related the position 1 to 6 in (c), and the electron beam size was about 0.5 nm under STEM mode. We detected the particles of 1, 2 and 4 as well as at position 3 (the edge of particle) and the composition were In, Sn and O and Si. It showed predominantly Si and O at position 5 between two particles, and In, Sn and O at the position 6 in the ITO film. The EDS detections showed the particles are ITO particles that are immerged in the SiO$_2$ substrate. This could be related to the XRD results of the modification at the interface of ITO/SiO$_2$ after AO exposure.

We found that ITO films were modified after exposure to 5 eV atomic oxygen. In previous studies of ground simulation of AO, Synowicki *et.al* and Woollam *et.al.* reported less reaction

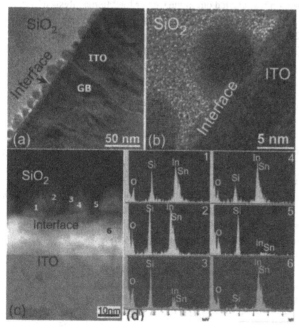

Figure 4. (a) Cross-sectional TEM image of ITO film, (b) HRTEM at the interface of ITO/SiO$_2$, and (c) HAADF image as a reference for EDS spectra as shown in (d), where spectra 1 to 6 were related the position 1 to 6 in (c), after exposed to AO with fluence: 2×10^{20} atom/cm^2.

with a thermal asher AO source[7,8]. However, ITO coatings were highly modified when they were exposed to AO on the ram side of the space vehicle in the LEO due to the 5 eV kinetic energy [7,8]. The nano-particles of ITO formed near the interface in the SiO_2 substrate reveals the AO reaction is at the interface. The non-contacting grain boundaries in ITO films can supply the pathway for the AO transport across the ITO film to the interface. This may be detrimental to the application of ITO coating to protect polymers from AO erosion in the low Earth orbital environment. It is reported that the oxygen deficiency in ITO film is common[6,11-13]. We speculate that the O-deficient ITO films were oxidized and the volume may be expanded under the AO reaction. The ITO particles could be pushed into SiO_2 by the volume expansion. The volume expansion may also increase the surface roughness of the films.

CONCLUSION

The 5 eV AO modified the ITO surfaces by increasing the surface roughness with the increasing AO fluences. Many ITO particles with diameters from 2 to 10 nm were created in the SiO_2 substrate near the interface after AO exposure, revealing the O atoms travel through the ITO film to the interface. The grain boundaries of the columnar grown ITO film may supply pathway for the AO transport.

ACKNOWLEDGMENT

This Multi-University Research Initiative (MURI) program is supported through Air Force Office of Scientific Research under grant F49620-01-1-0336 by Department of Defense (DoD). Some of the structural characterization was carried out in the Center for Microanalysis of Materials, University of Illinois, which is partially supported by the U.S. Department of Energy under grant DEFG02-96-ER45439. We would like to thank L. Wang and Dr. J.G. Wen at University of Illinois at Urbana-Champaign and A. M. Stewart, Prof. J. Leonard, M. Kisa and Cole Van Ormer at University of Pittsburgh for their assistance and discussion.

REFERENCES

1. A. E. Hedin, Journal of Geophysical Research, [Space Physics] 92, 4649 (1987).
2. M. R. Reddy, Journal of Materials Science 30, 281 (1995).
3. F. L. Bouquet and C. R. Maag, IEEE Transactions on Nuclear Science NS-33, 1408 (1986).
4. A. R. Chambers, I. L. Harris, and G. T. Roberts, Materials Letters 26, 121 (1996).
5. T. K. Minton and D. J. Garton, in Chemical Dynamics in Extreme Environments, edited by R. A. Dressler (World Scientific, Singapore, 2001).
6. M. R. Reddy, N. Srinivasamurthy, and B. L. Agrawal, Surface and Coatings Technology 58, 1 (1993).
7. R. A. Synowicki et.al., Surface and Coatings Technology 62, 499 (1993).
8. J. A. Woollam et.al., Thin Solid Films 241, 218 (1994).
9. G. E. Caledonia, R. H. Krech, and B. D. Green, AIAA Journal 25, 59 (1987).
10. S. K. Rutledge et.al., J Am Inst Conserv 39, 65 (2000).
11. K. L. Chopra, S. Major, and D. K. Pandya, Thin Solid Films 102, 1 (1983).
12. J. C. C. Fan and J. B. Goodenough, Journal of Applied Physics 48, 3524 (1977).
13. R. X. Wang et.al., Journal of Applied Physics 97, 033504(2005).

Mater. Res. Soc. Symp. Proc. Vol. 887 © 2006 Materials Research Society　　　　0887-Q06-04

Degradation of Assembled Silicon Nanostructured Thin Films: A Theoretical and Experimental Study

Valeria Bertani, Luisa D'Urso, Alessandro A. Scalisi, Giuseppe Compagnini and Orazio Puglisi
Dipartimento di Scienze Chimiche, Università di Catania, Viale A. Doria 6 – 95125 Catania
(ITALY)

ABSTRACT

The study of the structures and properties of small elemental clusters has been an extremely active area of current research, due to the peculiar behavior of these species halfway between that of single atoms and of the bulk phase. In this work silicon nanoclusters are generated by ablation of a high purity polycrystalline rod with a pulsed laser vaporization source and then deposited on a support. Their structure is studied both in the gas phase by means of Time of Flight Mass Spectrometry and in the solid phase through *in situ* Raman and Infrared Spectroscopy. The spectra reveal that the as deposited clusters are hydrogenated with negligible amount of oxide. Degradation of silicon nanoclusters has been studied after gas exposure. In the gas of air a consistent modification was observed, leading to a near-infrared luminescent silicon nanoparticles. In the second part of the work, density functional theory is applied to investigate the geometrical structure of silicon clusters and their interaction, in term of structure and energy, with different gases. The calculations were performed with the Gaussian 03 program suite, adopting the B3LYP functional to calculate the exchange and correlation energy. Si_8 has been chosen as model cluster to study the degradation of silicon clusters both kinetically and thermodynamically, in order to explain the experimental evidences. Experimental and calculated infrared spectra are compared.

INTRODUCTION

In these last years a lot of work has been done on small silicon nanoparticles, both experimentally and theoretically, because of their peculiar reactivity and properties which make them useful for different applications. On the other hand, this high ratio surface/volume is responsible for cluster degradation when exposed to external agents. As an example, thin film hydrogenate amorphous silicon is a technically and economically-feasible material for solar cells but, due to the presence of hydrogen atoms, as it results from molecular dynamics calculations, it undergoes upon light-induced degradation [1]. An useful way to prevent the degrade of silicon nanoparticles, especially of their photoluminescence, is the creation of a protective layer on their surface by embedding them in a polymer, or by refluxing a silicon nanoparticle dispersion in the presence of an alkene [2] or by means of a UV-induced graft polymerization of acrylic acid on particle surface [3].

In the present study, degradation of silicon nanoparticles, produced by Low Energy Cluster Beam Deposition, was investigated both experimentally and theoretically. In particular, reactions of small silicon clusters with hydrogen and oxygen were studied.

EXPERIMENTAL DETAILS

Silicon clusters were obtained by ablation of a high purity (99.9999 %) polycrystalline silicon rod with a pulsed laser vaporization source (with a Nd:YAG laser operating at 532 nm and 10 Hz repetition rate) in a pulsed flow of helium buffer gas (5 bar). Clusters left the chamber through an expansion nozzle with a supersonic motion, due to the pressure gradient (at the exit of the source pressure is maintained at about 10^{-7} mbar). The supersonic beam was then directed to a Time of Flight mass spectrometer, situated orthogonally to the propagation direction, or deposited on a substrate positioned at a distance of 0.8 m from the expansion nozzle. LECBD technique permits to obtain nano-structured materials with peculiar structure and properties because the generated clusters are not fragmented when they land on the substrate, having low energies (a few hundreds of meV per atom), as it is shown in Figure 1 [4,5]. In order to investigate oxygen-induced degradation, the deposited silicon aggregates were analyzed both into the deposition vacuum chamber and in air (at atmospheric pressure) with Raman and FT-IR spectroscopies [6,7]. Depending on the analytic technique to be used, Highly Oriented Pyrolytic Graphite (HOPG), silicon wafers or KBr pellets were chosen as substrates.

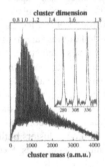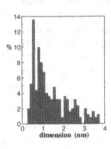

Figure 1. TOF-MS spectrum of Si_n anions (on the left) and Si_n distribution in the deposited film, as measured with AFM.

CALCULATIONS

The geometrical structure and stability of silicon clusters (up to 10 atoms) and their interaction, in term of structure and energy, with hydrogen and oxygen were studied with Density Functional Theory (DFT), having the advantage over other more sophisticated *ab initio* methods of scaling with the square or cube of the problem size. The stability of all the optimized structures was verified computing the force constants and vibrational frequencies. All calculations were performed adopting the three-parameter Becke Lee-Yang-Parr hybrid (B3LYP) functional to determine the exchange and correlation energies [8] and 6-31G [9], with the addition of single first polarization functions, as basis set. Gaussian 03 was adopted as program suite [10].

A variety of theoretical works on silicon nanoparticles are present in recent literature, due to the high interest in Si-based devices and at the same time to the relatively small number of electrons of this atom, which makes possible calculations, also of big species. Different computational techniques (e.g. empirical potentials [11], Hartree-Fock [12], MP2 [13-15], molecular dynamics [16]),have been used to investigate neutral as well as ionic silicon clusters. Moreover, in the DFT framework, different combinations of functionals and basis sets have been adopted [17-21].

DISCUSSION

Calculations on structures and cohesive energies of silicon clusters, both for neutral and ionic species, were first performed. By an inspection of Figure 2 it comes out that differently charged particles possess similar energies (in particular $\Delta E(Si_n^-) > \Delta E(Si_n^+) > \Delta E(Si_n)$); in all cases, stability increases up to n = 6 while for bigger clusters is more or less the same. Structures are similar too, as it is shown by the two examples reported and are in good agreement with literature [12,15,17-23].

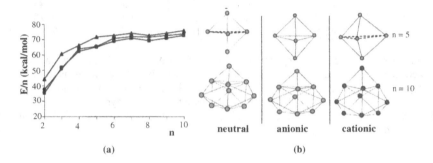

(a)   (b)

Figure 2. (a) Average bond energies per atom (kcal/mol) as a function of cluster dimension n for Si_n (square), Si_n^+ (circle) and Si_n^- (triangle); (b) Structures of neutral and ionic clusters with 5 and 10 Si atoms.

Figure 3 reports Infrared and Raman spectra of the nanostructured thin film in vacuum and at air exposure.

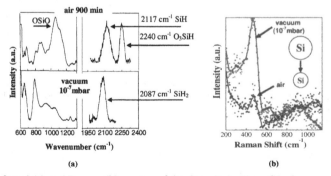

(a)   (b)

Figure 3. Infrared (a) and Raman (b) spectra of the deposited silicon film in vacuum and at air exposure.

Silicon nanoparticles are hydrogenated already at the deposition chamber pressure of 10^{-6} mbar (clusters react with the small amount of impurities of the system), so, to mime the experimental results, the interaction of silicon clusters with hydrogen was then considered in DFT calculations. Hydrogen bond energies and Si-H stretching frequencies of the most stable structures were calculated and results are reported in Table I. Clusters with $n \geq 8$ have comparable behaviors, so Si_8 was chosen as model cluster to study all further reactions.

Table I. Calculated hydrogen bond energies (BE, kcal/mol) and Si-H stretching frequencies (v, $^{cm-1}$) in Si_nH clusters (n = 1-10).

n	BE (kcal/mol)	v (cm^{-1})	n	BE (kcal/mol)	v (cm^{-1})
1	74.2	2022	6	53.4	2181
2	70.1	1480	7	33.8	2054
3	47.5	1304	8	48.4	2144
4	37.6	1489	9	46.6	2176
5	62.4	1315	10	48.0	2167

Left hand side of Figure 4 reports the reactions of Si_8 (**A**) with hydrogen: two molecules of H_2 were considered, leading to the formation of **C**, characterized by two adjacent di-hydrogenated Si atoms, as the experimental Si-H stretching frequency in vacuum (2087 cm^{-1}) is indicative of Si-H_2 species. Both steps of hydrogenation are thermodynamically favored, with a global energetic gain of 33 kcal/mol. The calculated Si-H stretching frequency is 2173 cm^{-1}, in good agreement with the experimental one.

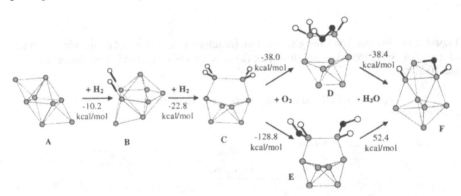

Figure 4. A scheme of hypothesized hydrogenation ad oxidation reactions on Si_8. Energies involved in each step are reported. Oxygen and hydrogen atoms are represented in black and with, respectively.

Oxygen degradation was then considered, starting from the di-hydrogenated silicon cluster **C**. The investigated global reaction is the addition of a molecule of O_2 and the subsequent loss of a molecule of water, with an energetic gain of 76.4 kcal/mol (right hand side of Figure 3). The final product (**F**) shows H-Si-Si and H-Si-O structures, in agreement with experimental IR spectrum (v_{exp}(Si-Si—H)$_{stretching}$ = 2117 cm^{-1} vs. v_{calc}(Si-Si—H)$_{stretching}$ = 2157 cm^{-1}; v_{exp}(O-Si—H)$_{stretching}$ = 2240 cm^{-1} vs. v_{calc}(O-Si—H)$_{stretching}$ = 2224 cm^{-1}). Two different intermediates for oxygen addition were supposed, respectively **D** and **E** in the scheme, but the pathway through **E** results not favored by thermodynamics.

CONCLUSIONS

The reactions of silicon clusters with hydrogen and oxygen were studied both experimentally and theoretically. Nanostructured thin films were prepared by LECBD and characterized in vacuum and in air with Infrared and Raman spectroscopies. In order to better understand degradation mechanisms, DFT calculations on Si_n clusters were performed obtaining results in good agreement with experimental data. Oxidation needs a more detailed study: different mechanisms should be considered together with the kinetics of elementary reactions. Moreover, the addition of other oxygen molecules, and consequently the formation of a more oxidized product, should be considered in order to explain the experimental infrared bands typical of O-Si-O structures.

ACKNOWLEDGMENTS

Authors gratefully acknowledge MIUR (C26P14A) for the financial support.

REFERENCES

1. Q. Li and R. Biswas, *Phys. Rev. B* **52**, 10705 (1995).
2. W. D. Kirkey, A. N. Cartwright, X. Li, Y. He, M. T. Swihart, Y. Sahoo and P. N. Prasad in Quantum Dots, Nanoparticles and Nanowires, edited by P. Guyot-Sionnest, H. Mattoussi. U. Woggon and Z.-L. Wang (Mater. Res. Soc. Symp. Proc. 789, Pittsburgh, PA, 2004).
3. Z. F. Li and E. Ruckenstein, *Nano Letters* **4**, 1463 (2004).
4. J. E. Bower and M. F. Jarrold, *J. Chem. Phys.* **97**, 8312 (1992).
5. P. Melinon, P. Keghelian, B. Prevel, V. Dupuis, A. Perez, B. Champagnon, Y. Guyot, M. Pellarin, J. Lermé, M. Broyer, J. L. Rousset and P. Delichere, *J. Chem. Phys.* **108**, 4607 (1998).
6. G. Compagnini, L. D'Urso, A. A. Scalisi, O. Puglisi and B. Pignataro, in publication on *Thin Solid Films* (2005).
7. G. Compagnini, L. D'Urso and O. Puglisi, in publication on *Mat. Sci. & Eng. C* (2005).
8. (a) A. D. Becke, *J. Chem. Phys.* **98**, 5648 (1993). (b) C. Lee, W. Yang and R. G. Parr, *Phys. Rev. B* **37**, 785 (1988).
9. (a) R. Ditchfield, W. J. Hehre, and J. A. Pople, *J. Chem. Phys.* **54**, 724 (1971). (b) W. J. Hehre, R. Ditchfield, and J. A. Pople, *J. Chem. Phys.* **56**, 2257 (1972). (c) P. C. Hariharan

and J. A. Pople, *Mol. Phys.* **27**, 209 (1974). (d) M. S. Gordon, *Chem. Phys. Lett.* **76**, 163 (1980). (e) P. C. Hariharan and J. A. Pople, *Theo. Chim. Acta* **28**, 213 (1973). (f) J.-P. Blaudeau, M. P. McGrath, L. A. Curtiss, and L. Radom, *J. Chem. Phys.* **107**, 5016 (1997). (g) M. M. Francl, W. J. Pietro, W. J. Hehre, J. S. Binkley, D. J. DeFrees, J. A. Pople, and M. S. Gordon, *J. Chem. Phys.* **77**, 3654 (1982). (e) R. C. Binning Jr. and L. A. Curtiss, *J. Comp. Chem.* **11**, 1206 (1990). (e) V. A. Rassolov, J. A. Pople, M. A. Ratner, and T. L. Windus, *J. Chem. Phys.* **109**, 1223 (1998). (f) V. A. Rassolov, M. A. Ratner, J. A. Pople, P. C. Redfern, and L. A. Curtiss, *J. Comp. Chem.* **22**, 976 (2001).

10. Gaussian 03, Revision C.01, M. J. Frisch, G. W. Trucks, H. B. Schlegel, G. E. Scuseria, M. A. Robb, J. R. Cheeseman, J. A. Montgomery, Jr., T. Vreven, K. N. Kudin, J. C. Burant, J. M. Millam, S. S. Iyengar, J. Tomasi, V. Barone, B. Mennucci, M. Cossi, G. Scalmani, N. Rega, G. A. Petersson, H. Nakatsuji, M. Hada, M. Ehara, K. Toyota, R. Fukuda, J. Hasegawa, M. Ishida, T. Nakajima, Y. Honda, O. Kitao, H. Nakai, M. Klene, X. Li, J. E. Knox, H. P. Hratchian, J. B. Cross, V. Bakken, C. Adamo, J. Jaramillo, R. Gomperts, R. E. Stratmann, O. Yazyev, A. J. Austin, R. Cammi, C. Pomelli, J. W. Ochterski, P. Y. Ayala, K. Morokuma, G. A. Voth, P. Salvador, J. J. Dannenberg, V. G. Zakrzewski, S. Dapprich, A. D. Daniels, M. C. Strain, O. Farkas, D. K. Malick, A. D. Rabuck, K. Raghavachari, J. B. Foresman, J. V. Ortiz, Q. Cui, A. G. Baboul, S. Clifford, J. Cioslowski, B. B. Stefanov, G. Liu, A. Liashenko, P. Piskorz, I. Komaromi, R. L. Martin, D. J. Fox, T. Keith, M. A. Al-Laham, C. Y. Peng, A. Nanayakkara, M. Challacombe, P. M. W. Gill, B. Johnson, W. Chen, M. W. Wong, C. Gonzalez, and J. A. Pople, Gaussian, Inc., Wallingford CT, 2004.

11. S. Yoo and X. C. Zeng, *J. Chem. Phys.* **119**, 1442 (2003).

12. K. Ragavachari, C. M. Rohlfing *J. Chem. Phys.* **89**, 2219 (1988).

13. C. Majumder and S. K. Kulshreshtha, *Phys. Rev. B* **69**, 115432 (2004).

14. C. Xu, T. R. Taylor, G. R. Burton and D. M. Neumark, *J. Chem. Phys.* **108**, 1395 (1998).

15. C. Zhao and K. Balasubramanian, *J. Chem. Phys.* **116**, 3690 (2002).

16. P. Ballone W. Andreoni, R. Car and M. Parrinello, *Phys. Rev. Lett.* **60**, 271 (1988).

17. V. E. Bazterra, M. C. Caputo, M. Ferraro and P. Fuentealba, *J. Chem. Phys.* **117**, 11158 (2002).

18. S. Nigam, C. Majumder and S. K. Kulshreshtha, *J. Chem. Phys.* **121**, 7756 (2004).

19. R. Fournier, S. B. Sinnott and A. E. DePristo, *J. Chem. Phys.* **97**, 4149 (1992).

20. A. A. Shvartsburg, B, Liu, M. F. Jarrold and K.-M. Ho, *J. Chem. Phys.* **112**, 4517 (2000).

21. K. Raghavachari and C. M. Rohlfing, *J. Chem. Phys.* **94**, 3670 (1991).

22. M. C. McCarthy and P. Thaddeus, *Phys. Rev. Lett.* **90**, 213003 (2003).

23. X. Zhu and X. C. Zeng, *J. Chem. Phys.* **118**, 3558 (2003).

Mater. Res. Soc. Symp. Proc. Vol. 887 © 2006 Materials Research Society　　　0887-Q08-03

Statistical Mechanics: Undamaged ⇐-radiation-⇒ Damaged Atomic Lattice Density Evolution and Stochastic Deformation Functional

Ray B. Stout[1] and Natasha K. Stout[2]
[1]RhoBetaSigma Affaires, Livermore, CA 94550
[2]Center of Risk Analysis, Harvard School of Public Health, Boston, MA

Abstract The deformation kinematics for radiation damage response of bulk materials is presently semi-empirical and phenomenological based on a continuum mechanics supposition: *there exists a function space of continuous functions to describe material displacement, strain, strain-rate metrics by using the mathematics of differential calculus.* Existing data being assembled from tests on nano-length-scale samples provide objective evidence that the continuum mechanics supposition is not an adequate generic mathematical description for radiation damage response in surface-dominated material nano-structures. An alternative approach will be described that uses concepts and methods from classical statistical mechanics and describe deformation kinematics as a stochastic accumulation of discrete damage events at atomic lattice nano-length-scales. Although radiation damage deformation at a lattice length-scale in solids is mechanistically different from velocity scattering developed by Boltzmann for a kinetic theory of gases, the two problem areas are technically similar and in some simply cases there are useful mathematical analogs. The technical similarities and mathematical analogs will be used to define stochastic density functions (number per unit volume functions) for undamaged and damaged "size and size change" lattice species (similar to a probabilistic density function for atomic velocities in gas theory). In general, equations for undamaged and damaged lattice density function evolution are Boltzmann-type equations, which can be approximated and solved for simple cases of radiation induced material damage. Using a path integral approach and the two stochastic density functions, a stochastic functional will be derived for the relative deformation between any two arbitrary spatial points in a radiation damaged material. Given the relative deformation as an explicit functional of the undamaged and damaged lattice density functions, the kinematics metrics of relative velocity, strain, strain rate, etc., will also be functionals. In the case of radiation damage and annealing, the two lattice density functions are analog expressions to those commonly used to model "birth and death" population evolution.

Introduction Radiation damage research of materials has over a fifty year history[1]. The deformation kinematics for damage response of bulk materials is semi-empirical and phenomenological based on a continuum mechanics supposition. The supposition assumes that there exists a function space of continuous functions to describe material displacement, strain, strain-rate metrics by using the mathematics of differential calculus. Existing data assembled from tests on nano-length-scale(NLS) samples provide objective evidence that this continuum mechanics supposition is not an adequate generic mathematical description of damage response for surface-dominated material structures at NLS. Recent radiation damage analysis[2] used classical statistical mechanics concepts to define stochastic density functions for the number density of undamaged, $N(\underline{x},t; \underline{q})$, and the number density of damaged, $D(\underline{x},t; \underline{q})$, atomic lattices at spatial point $\underline{x}$ and lattice species $\underline{q}$ per unit spatial volume $d\underline{x}$ and per unit lattice species volume $d\underline{q}$ at time t. Boltzmann[3] type evolution equations exist to describe the interdependent evolution of these two lattice density functions during radiation damage and damage annealing kinetics; as species of the $N(\underline{x},t; \underline{q})$ and $D(\underline{x},t; \underline{q})$ lattice density functions are destroyed and created by these two physical processes. The $\underline{q}$ lattice species variable contains physical attribute vector variables for the length dimensions of a lattice cell and time rates of the cell length dimensions for any lattice cell species. A relative deformation functional between two arbitrary spatial points is derived as a path integral functional. The integral functional depends on the stochastic lattice density functions $N(\underline{x},t; \underline{q})$ and $D(\underline{x},t; \underline{q})$. Using integral calculus mathematics[4, 5, 6], path integrations can be evaluated over a countable stochastic set of arbitrary spatial paths. For a contiguous material, the vector measure of relative deformation exists as an invariant value independent of the physical spatial path between the two arbitrary spatial points. This path integral invariance is an analytical basis to statistically derive decomposition metrics for undamaged and damaged spatial sub-domains between any two arbitrary points. Variations in the spatial vector between two arbitrary points, by varying the position of one of the two points relative to the other, is an analytical basis to statistically derive length-scale metrics that are not in continuum mechanics models. These analyses address stochastic radiation damage evolution and will begin to illuminate physical enigmas and gray, shadowy representations of continuum mechanics at nano-length-scales. The stochastic models of radiation damage and annealing can be described in some cases as analogs to those commonly used in "classical birth and death population" models [7, 8] in biological and social sciences.

Background The multi-collision atomic events from scattering down in velocity(or momentum) of high energy Mev charged or non-charged radiation particles traveling through solids leaves behind a path of broken atomic bonds and displaced atomic structures [1, 2, 9]. In crystalline solids, the displaced atomic structures are vacancy sites, interstitial atomic sites, and substitutional atomic sites; which can evolve into line defect structures identified as dislocation species and into free surface structures identified as voids and bubbles. In amorphous, organic, and biological solids, similar atomic structure mechanisms are expected to evolve. However, in these cases one does not typically see discussions of dislocation evolution. But in organic materials, there are hydro-carbon and hydrogen gas releases. Most all studies with observations at or near atomic length scales are in single bulk materials. Some interface and embedded particle observations show radiation induced transport and damage in spatial neighborhoods of the interior interfaces and surfaces[9]. The lattice density function analysis and continuum approximation problems are complex and partially complete for mathematical representations that describe atomic structural evolution and deformations for just the single, bulk material cases. For inhomogeneous materials with interior interfaces and surfaces typical of nano-scale gages as illustrated in Fig. 1, the lattice density function analysis are more complex and less complete for representations describing atomic structural evolution and deformations from radiation, as well as induced transport and damage across interior surfaces into spatial neighborhoods of the adjacent material. In either case, a primary objective of this paper is to develop a lattice density approach that can replace the classical concept of using the total number of displaced atoms as a metric to correlate data from radiation damage.

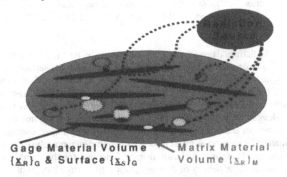

Gage Material Volume   Matrix Material
$\{\underline{x}_R\}_G$ & Surface $\{\underline{x}_S\}_G$ Volume $\{\underline{x}_R\}_M$

Figure 1. Illustration of radiation damage events in a nano-scale gage material.

The total number of displaced atoms is a scalar metric. As such, a scalar metric is physically and mathematically inadequate to determine and prescribe general materials' response models. Particularly when there exist a significant dependence on lattice density deformations and/or radiation induced atomic transport across interior materials' surfaces. In these cases, deformations metrics are displacement vectors and strains (a second order tensor) and transport across a surface is a flux vector. For engineering applications, neither vectors nor strain material responses can be physically or mathematically determined by a scalar; except perhaps for strictly isotopic, homogeneous conditions which are precluded by arbitrary interior surfaces.

In the next section, a construction for undamaged lattice density functions is developed by associating with each material sub-element a local sub-volume, on a length scale of atomic or molecular dimensions on the order of ~10^{-10}m. For each spatial sub-volume, a Cartesian vector spatial domain is assigned to the material sub-element; for example three lattice vectors ($\underline{a}$, $\underline{b}$, $\underline{c}$). In the case of a crystalline lattice structure for a solid, the three vectors are locally repeated in regular and translational congruent operations throughout a stress-free grain volume, and the three vectors will have a rotated orientation from grain to grain. In crystalline solids, this is short range ordering of the lattice structure vectors. For amorphous materials, or for organic and biological materials with primarily carbon-hydrogen molecules, the assignment of three "lattice structure" vectors to approximate the sub-volume of an enclosed and repeating multi-atom structure can be defined. However, the conceptualization of locally repeated regular and translational congruent operations of the three "lattice structure" vectors throughout a stress-free volume is not correct, as the regular and translational congruence of short range ordering amongst spatially adjacent lattice structure vectors is not necessarily physically valid. Nonetheless, a statistical mechanics description of lattice structure sub-volumes can be proposed for amorphous, organic, and biological materials; and a lattice density function defined for this more general case. The mathematical constructs for undamaged lattice density functions of amorphous, organic, and biological materials are analogs to those defined in the simpler conceptual case for crystalline solids. Once density functions are defined, generic Boltzmann balance equations, which are stochastic accounting statements, can be written to describe the evolution of the lattice density functions due to radiation effects.

Given an initial, undamaged lattice density function, the primary phenomenological processes of an impinging radiation flux are to transform undamaged lattice sites into damaged lattice sites along the scattering down path of a high-energy radiation particle. The rates at which the damaged lattice sites are created are functional terms in the Boltzmann balance equation. Thus, radiation scattering processes create a density of damaged lattice sites; these damaged lattice sites are a complementary lattice density to the undamaged lattice density function. In the simplest case for single, bulk material models, the interior spatial regions will have a net conservation for the sum of undamaged and damaged lattice sites, as an undamaged lattice site, containing an atom or molecular structure, can presumably only become a damaged lattice site[2]. However, in more complex cases, as mentioned above, there are several species, or types, of damaged lattices sites that can be identified; such as vacancy sites, interstitial atomic sites, and substitutional atomic sites. In the following, these will not be explicitly modelled as distinct lattice density functions, but will be aggregated as discrete lattice species in the damaged lattice density function. This simplification will provide a greatly shortened description for undamaged and damaged lattice deformation functionals.

Preliminary: Undamaged to Damaged Lattice Evolution and Deformations

An undamaged lattice cell structure for crystalline solids is described as a highly spatially ordered set of "small rectangular boxes" regularly stacked in a material volume. For an idealized crystalline material, each rectangular box is an idealized lattice cell that contains an atom or a molecule. In the case of amorphous or organic solids, the undamaged lattice cell structure is described as a low spatially ordered set of "small non-rectangular boxes" that are not regular stacked in a material volume. When the material volume has no open spatial point sets (interior surfaces enclosing holes or bubble structures), then the material volume is termed contiguous. Typical linear dimensions of a lattice cell are about $\sim10^{-10}$m. Thus, per cubic centimeter, there are about $\sim10^{+24}$ lattice cells. As will be discussed below, the relative vector separation between two spatially separated lattice cells can be evaluated by simply adding the vectors of the "boxes" along an arbitrary path between the lattice cells. To fully describe three dimensional deformations by spatial integrations over a density of lattice cells, each lattice cell species will require a three-vector set of dimensions. Therefore, dimensions for the spatial size of a rectangular lattice cell species are denoted with three vector attributes ($\underline{a}$, $\underline{b}$, $\underline{c}$) whose tensor component values (a_i, b_i, c_i) are with respect to an arbitrary Cartesian coordinate reference frame. Thus, in order to describe relative deformation between any two arbitrary spatial points, say point B relative to point A, with the formalism of statistical mechanics, the undamaged density function N and damaged density function D will both depend explicitly on the physical attribute variables of lattice cell species vectors ($\underline{a}$, $\underline{b}$, $\underline{c}$). From this definition, and in an un-deformed crystalline lattice structure at time t, a relative position vector from a spatial point at $\underline{x}_A$ to a spatial point at $\underline{x}_B$ is a relative position vector denoted by $\underline{X}(A \Rightarrow B, t)$ between lattice cells at points A and B. For a set of idealized, non-deformed lattice cells in a crystalline material, the attribute vector $\underline{a}$ can be aligned with the arbitrary Cartesian coordinate vector $\underline{X}_1$ direction, attribute vector $\underline{b}$ can be aligned with the arbitrary Cartesian coordinate vector $\underline{X}_2$ direction, and attribute vector $\underline{c}$ can be aligned with the arbitrary Cartesian coordinate vector $\underline{X}_3$ direction. In this un-deformed lattice case, the relative position vector between spatial positions A to B is represented as the vector sum of the number of lattice cells in each of the three directions, say n_{x1}, n_{x2}, and n_{x3} times the lattice cell vector dimension in that direction, simply as:

$$\underline{X}(A \Rightarrow B, t) = n_{x1} \underline{a} + n_{x2} \underline{b} + n_{x3} \underline{c} \tag{1}$$

In this idealized case, the relative position vector $\underline{X}(A \Rightarrow B, t)$ is the diagonal vector of a rectangular box between points A and B, and it exists as an invariant independent of the spatial path along which one counts the "net number" (n_{x1}, n_{x2}, n_{x3}) of lattice cell vectors between points A and B. This suggests that Eq(1) can be written more generally as a path independent functional that depends on the lattice density functions. The path integral functional form of Eq(1), that depends on the lattice number density functions $\{N(\underline{x}, t; \underline{a}, \underline{b}, \underline{c}) + D(\underline{x}, t; \underline{a}, \underline{b}, \underline{c})\}d(vol(\underline{x}))d\underline{a}d\underline{b}d\underline{c}$ (this is the number of (a_i, b_i, c_i) undamaged and damaged lattice cell species in a spatial volume element $d(vol(\underline{x}))$) and lattice species volume element $d\underline{a}d\underline{b}d\underline{c}$ at point $\underline{x}$ and time t is expressed by writing the spatial volume element $d(vol(\underline{x}))$ in terms of local lattice vector attributes for the terms in the integrand. For the first term, the local spatial volume element $d(vol(\underline{x}))$ is $E_{pmn}b_m c_n dx_p$ (here the third order alternating Cartesian tensor, E_{pmn}, is used to write a volume element of integration) for the $\underline{a}$ vector direction integrand, for the second term the spatial volume element $d(vol(\underline{x}))$ as $E_{pmn}c_m a_n dx_p$ for the $\underline{b}$ vector direction integrand, and for the third term the spatial volume element $d(vol(\underline{x}))$ is $E_{pmn}a_m b_n dx_p$ for the $\underline{c}$ vector direction integrand. Thus, Eq(1) can be written as a path integral functional that sums the number of $\underline{a}$, $\underline{b}$, and $\underline{c}$ vector lattice units in the kth component direction of vector $X_k(A \Rightarrow B, t)$ between points A and B as

$$X_k(A \Rightarrow B, t) = \int_{(A \Rightarrow B, t)} \int_{(\underline{a}, \underline{b}, \underline{c})} [a_k E_{pmn}b_m c_n + b_k E_{pmn}c_m a_n$$
$$+ c_k E_{pmn}a_m b_n] \{N(\underline{x}, t; \underline{a}, \underline{b}, \underline{c}) + D(\underline{x}, t; \underline{a}, \underline{b}, \underline{c})\}d\underline{a}d\underline{b}d\underline{c} \, dx_p \tag{2}$$

The integration in Eq(2) to evaluate the relative position vector $X_k(A \Rightarrow B, t)$ is a general path independent calculus given any probability density functions $N(\underline{x}, t; \underline{a}, \underline{b}, \underline{c})$ and $D(\underline{x}, t; \underline{a}, \underline{b}, \underline{c})$ for lattice cell density, whether the lattice cells of the density function are deformed or un-deformed. A path independence integral for relative material positions and motions are fundamental to mathematical descriptions of physical space-time model development [5]. Given this, it is clear that the present form of Eq(2) does not describe the evolution of deformation during an arbitrary time interval as a functional of the evolution of both undamaged and damaged lattice densities N and D. To describe the deformation evolution, the integral dependence on the microscopic spatial time rates of deformation for these two density functions must be included as part of the path integration calculus. The deformation response will be completed in a following section, but a preliminary discussion is necessary here to describe evolution equations for densities N and D. There are several contributions to micro-scale, lattice deformations: first there are "undamaged to damaged" finite lattice size changes as radiation particles create sub-volumes lattice species; second, there are "damaged to undamaged" finite lattice size changes during annealing; and third, there are time continuous, elastic lattice size changes at finite rates in each of the lattice vectors.

The first contribution is from lattice cells with size vectors ($\underline{a}$, $\underline{b}$, $\underline{c}$) in the undamaged lattice density N becoming damaged lattice cells in density D of size ($\underline{a}+\Delta\underline{a}$, $\underline{b}+\Delta\underline{b}$, $\underline{c}+\Delta\underline{c}$). Here, the lattice vector increments, ($\Delta\underline{a}$, $\Delta\underline{b}$, $\Delta\underline{c}$) = ($\Delta\underline{q}$), are approximated as instantaneously occurring micro-scale changes (during a high energy collision radiation-atomic atom scattering event on a finite infinitesimal time scale interval Δt_{Rad}) in lattice cell species size as transitions occur between undamaged to damaged species. This will be modeled as a dependence on the rate of decrease in undamaged lattice species in the density rate function $\partial N(\underline{x}, t; \underline{a}, \underline{b}, \underline{c})/\partial t$ versus a corresponding increase in damaged lattice species in the damaged density function rate represented as a differential rate expression: $\partial D(\underline{x}, t; \underline{a}+\Delta\underline{a}, \underline{b}+\Delta\underline{b}, \underline{c}+\Delta\underline{c})/\partial t = \partial D(\underline{x}, t; \underline{q}+\Delta\underline{q})/\partial t$. The second contribution is from lattice cells with size vectors ($\underline{a}+\Delta\underline{a}$, $\underline{b}+\Delta\underline{b}$, $\underline{c}+\Delta\underline{c}$) in the damaged lattice density D annealing back to undamaged lattice cells in density N of size ($\underline{a}$, $\underline{b}$, $\underline{c}$). Here, the lattice vector increments, ($\Delta\underline{a}$, $\Delta\underline{b}$, $\Delta\underline{c}$) = ($\Delta\underline{q}$), are also approximated as instantaneously occurring micro-scale changes (during a atomic lattice annealing event on an infinitesimal time scale interval Δt_{Anneal}) in lattice cell species size as transitions occur from damaged to undamaged species. The annealing kinetics are modeled elsewhere[10] as a dependence on the rate of decrease in damaged lattice species in the density rate function $\partial D(\underline{x}, t; \underline{a}+\Delta\underline{a}, \underline{b}+\Delta\underline{b}, \underline{c}+\Delta\underline{c})/\partial t = \partial D(\underline{x}, t; \underline{q}+\Delta\underline{q})/\partial t$ versus a corresponding increase in undamaged lattice species in density function rate represented as a differential rate expression $\partial N(\underline{x}, t; \underline{q})/\partial t$.

The third contribution is from lattice cell dimensions changing at much slower velocities, and elasticity response requires that these rates be included as physical attribute variables of a lattice cell species. These rate attributes of a lattice cell species are denoted as ($\mathbb{a}_i$, $\mathbb{b}_i$, $\mathbb{c}_i$), where $\mathbb{a}_i$ is the time derivative of a_i, namely, da_i/dt, and similarly; $\mathbb{b}_i = db_i/dt$ and $\mathbb{c}_i = dc_i/dt$. This is notational lengthy, as the physical attributes of a lattice cell species are values in a point set whose domain is real numbers of the six vector space { a_i, b_i, c_i, $\mathbb{a}_i$, $\mathbb{b}_i$, $\mathbb{c}_i$ }. So for notational purposes, individual lattice species are denoted with variable $\underline{q}$, which is equal to (a_i, b_i, c_i, $\mathbb{a}_i$, $\mathbb{b}_i$, $\mathbb{c}_i$). Thus, $\underline{q}$ identifies the size and time rate of change in size of any lattice cell species, and for shorthand notational purposes in writing integrals the point set domain of lattice species is denoted by the set $\{\underline{q}\}$. This means that the density N with its variable arguments $N(\underline{x}, t: a_i, b_i, c_i, \mathbb{a}_i, \mathbb{b}_i, \mathbb{c}_i)$ is reduced to $N(\underline{x}, t; \underline{q})$. These lattice rate attributes must also be included in the density function $D(\underline{x}, t; \underline{q})$ of damaged lattice sub-volumes with its lattice species $\underline{q}$. Finally, the damaged lattice density function has one more physical attribute variable to model radiation dependent transport processes that occur from energetic collision events along the scatter-down radiation path[10]. Here, the undamaged to damaged transport effect will not be explicitly modeled; but it is similar to a statistical model for atomic diffusion velocity $v_{k[Rad]}$ relative to the local material velocity v_k.

Evolution Equations for Undamaged and Damaged Lattice Density Functions

In this section, evolution equations will be discussed for the undamaged lattice density, $N(\underline{x}, t; \underline{q})$, and the damaged lattice density, $D(\underline{x}, t; \underline{q})$. A Boltzmann transport equation for the undamaged lattice can be written for the term $N(\underline{x}, t; \underline{q})d\underline{q}dR(\underline{x})$, which is a function for the number of undamaged $\underline{q}$ species lattice cells in a $d\underline{q}$ species element that is in an isolated contiguous material volume element $dR(\underline{x})$ at time t. The transport equation describes the time rate of change of the function $N(\underline{x}, t; \underline{q})d\underline{q}dR(\underline{x})$ for all $\{\underline{q}\}$ species at all points in a spatial material volume $\{\underline{x}_R\}$ with spatial material surface $\{\underline{x}_S\}$. After simplifications, the Boltzmann transport equation for any "bulk" (no interior surface transport model) material radiation damage whether in the matrix volume $\{\underline{x}_R\}_M$ or gage spatial volume $\{\underline{x}_R\}_G$, is reduced to a population evolution equation:

$$\partial N(\underline{x}, t; \underline{q})/\partial t + \partial(v_k N(\underline{x}, t; \underline{q}))/\partial x_k + \underline{\underline{q}}\, \partial N(\underline{x}, t; \underline{q})/\partial q =$$

$$- \alpha_{k[Collision]} v_{k[Rado]} \mathbf{N}_{Rad}(\underline{x}, t; \underline{v}_{Rado})N(\underline{x}, t; \underline{q}) + \alpha_{anneal}\exp(-E_{anneal}/kT)D(\underline{x}, t; \underline{q}+\Delta\underline{q}) \qquad (3a)$$

A complementary Boltzmann transport equation for the damaged lattice can be written for the term $D(\underline{x}, t; \underline{q})d\underline{q}dR(\underline{x})$, which is a function for the number of damaged $\underline{q}$ species lattice cells in a $d\underline{q}$ species element that is in an isolated "bulk" contiguous material volume element $dR(\underline{x})$ at time t, and is:

$$\partial D(\underline{x}, t; \underline{q})/\partial t + \partial(v_k D(\underline{x}, t; \underline{q}))/\partial x_k + \underline{\underline{q}}\, \partial D(\underline{x}, t; \underline{q})/\partial q =$$

$$+ \alpha_{k[Collision]} v_{k[Rado]}\, \mathbf{N}_{Rad}(\underline{x}, t; \underline{v}_{Rado})N(\underline{x}, t; \underline{q} - \Delta\underline{q}) - \alpha_{anneal}\exp(-E_{anneal}/kT)D(\underline{x}, t; \underline{q}) \qquad (3b)$$

In Eqs(3), the undamaged and damaged lattice species do not contain an explicit description of radiation induced transport at a relative transport velocity $v_{k[Rad]}$ with respect to the local material velocity v_k. Also, the material velocity v_k is more complex than it appears and will be discussed as part of a deformation kinematics analysis in a following section. The gradient of material velocity times the density function is a nonlinear term in Eqs(3), and for small deformations and velocities will be neglected as they are second order nonlinear effects. Eqs(3) are accounting statements for a lattice species $\underline{q}$ in a spatial volume $\{\underline{x}_R\}$ with surface $\{\underline{x}_S\}$. For radiation damage in thin, nano-scale gage material volumes $\{\underline{x}_R\}_G$, the damage evolution Eq(3b) will typically require a damage lattice transport term across the gage material surface $\{\underline{x}_S\}_G$ as discussed below. The operational term, $\underline{\underline{q}}\, \partial\{...\}/\partial q$, describes elastic rate displacements in a Liouville lattice phase space, the term with coefficient $\alpha_{k[Collision]}$ describes the rate of destruction(death) of undamaged lattice species in Eq(3a) and the rate of creation(birth) of damaged lattice species in Eq(3b), and the term with coefficient α_{anneal} describes the rate of creation(birth) of undamaged lattice species in Eq(3a) and the rate of destruction(death) of damaged lattice species in Eq(3b). Physically, these are prescribed as discrete lattice events at atomic length scales for elastic length changes, for radiation damage length changes, and for annealing of radiation damage length changes in a dense lattice species phase space $\{\underline{q}\}$. Mathematically, these prescribed models use statistical mechanics' concepts to construct functional representations for their physical observable responses.

For the elastic evolution of the lattice density, the elasticity represents species transitions in undamaged lattice density as species $\underline{q}^*$ becomes a near neighbor species $\underline{q}$ and species $\underline{q}$ become a near neighbor species $\underline{q}^*$ in an arbitrarily small time interval dt[2]. For example, as a result of "time continuous" elastic deformations in a lattice cell size at $\underline{q}^*= (q^*, \underline{\underline{q}}^*)$, there is an increase in the neighboring lattice density species $\underline{q}$ where $q = q^* + \underline{\underline{q}}^*dt$ in an arbitrary dt time interval (here the rate of time change in lattice dimensional attributes, previous denoted as ($\underline{a}$, $\underline{b}$, $\underline{c}$), are written as $\underline{q}$). The elasticity will be "lattice conserving" for species as these elastic size changes do not increase or decrease the total number of lattice cells. Since the elastic deformation must be "integrate-able in the time domain without any path history dependence", the mathematical form for the elasticity operational term, $\underline{q}$ ∂..../∂q, is greatly restricted. And for the elastic transitions, it will be also required that the density functions N($\underline{x}$, t; $\underline{q}$) and D($\underline{x}$, t; $\underline{q}$) be invariant functions with respect to arbitrary translations in values for the size $\underline{q}$ attribute variable. For the elastic deformations, this means N($\underline{x}$, t - dt; $\underline{q}$) becomes equals to N($\underline{x}$, t ; $\underline{q}$ + $\underline{q}$dt, $\underline{c}$); as in this case lattice vector dimensions $\underline{q}$ are continuous in time for arbitrary, but finite, lattice rate vector $\underline{q}$ ($\underline{q}$ is integrate-able in time). The mathematically implication of the elastic operational term is the existence of a total time derivative, in the Liouville sense as illustrate in Fig 2. that is given by:

$$dN(\underline{x}, t; \underline{q})/dt = \partial N(\underline{x}, t; \underline{q})/\partial t + \underline{q}\, \partial N(\underline{x}, t; \underline{q})/\partial q \ \& \ \text{term } d\underline{q}/dt \, \partial N(\underline{x}, t; \underline{q})/\partial \underline{q} \text{ is } \sim 0 \qquad (4)$$

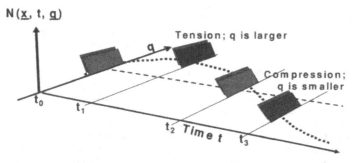

Figure 2. Elasticity lattice density changes "$\underline{q}$" to "$\underline{q} + \int_0^t \underline{q}dt$" in a Liouville phase space.

For the non-elastic lattice transitions, finite lattice size transformations $\underline{q}^* = \underline{q} + \Delta\underline{q}$ occur due to radiation events and due to lattice annealing events. For these first order radiation induced transformations, the rate of loss of undamaged lattice species is analytically, proportional to the existing undamaged lattice density, N($\underline{x}$, t; $\underline{q}$), times the collision cross-section vector, $\alpha_{k[Collision]}$ (which will in general depend on lattice attribute q, type of radiation, and radiation energy), multiplied by the radiation flux, $v_{k\,[Rado]}N_{Rad}(\underline{x}, t; v_{Rado})$. The time rate of change in undamaged lattice density, N($\underline{x}$, t; $\underline{q}$), of species $\underline{q}$ due to radiation is equal to the rate of increase in the damaged lattice density, D($\underline{x}$, t; $\underline{q}^*$), of species $\underline{q}^* = \underline{q} + \Delta q$. In models of stochastic population rates[7, 8], the loss rate for undamaged lattice species is a "death rate process" and the rate of increase for damaged lattice density is a "birth rate process". The size of the small length scale increments, $\Delta\underline{q}$, illustrated below in Fig 3 for undamaged to damaged lattice density and dimensional changes, can be effectively determined for phenomenological models from radiation experiments on bulk homogeneous materials. The experiments require measuring both deformation strain rates and strains simultaneously with radiation rate and fluence. In the following, the term accounting for annealing lattice transitions will be neglected to simplify and shorten expressions.

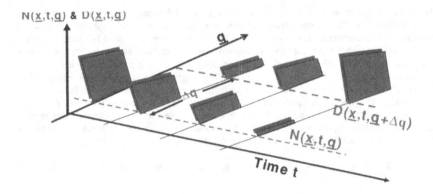

$N(\underline{x},t,\underline{q})$ & $D(\underline{x},t,\underline{q})$

$\underline{q}$

Δq

$D(\underline{x},t,\underline{q}+\Delta q)$

$N(\underline{x},t,\underline{q})$

Time t

Figure 3. Undamaged to damaged lattice species "$\underline{q}$" to "$\underline{q} + \Delta\underline{q}$" transitions due to radiation.

In bulk materials, Eqs(3) have approximations that neglect scatter-down path lengths and the number of the "collision scatter-down events"; as a high energy particle collision at spatial point $\underline{x}$ will create a damage lattice at $\underline{x}$, but may also will create and effectively transport a damaged lattice species to a local spatial vector set $\{\Delta\underline{X}\}$ in the neighborhood of spatial point $\underline{x}$. For bulk materials, the path lengths and number of events can be reasonably neglected as long as distances in set $\{\Delta\underline{X}\}$ are small compared to bulk material dimensions. For a nano-scale gage material embedded in a bulk material, the transport of damaged lattice species across gage surfaces is a potentially significant effect. The analytical details are addressed elsewhere [10] as a part of the evolution equation for damaged lattice density due to "collision scatter-down events" into a thin gage material. Here, a length scale ratio, $\{\Delta X_{k\ DM\text{-}max}\}n_k/\{\Delta X_{h\ G}\}n_h$, is used to approximate only matrix damaged lattice density transport into a thin, nano-scale material gage. This length scale ratio is the "collision scatter-down events" matrix material damage depth, $\{\Delta X_{k\ DM\text{-}max}\}n_k$, adjacent to and normal(unit normal vector n_k) to the gage material surface $\{\underline{x}_s\}_G$ divided by the nominal gage material thickness, $\{\Delta X_{h\ G}\}n_h$, normal(unit normal vector n_h) to the thin gage surface $\{\underline{x}_s\}_G$. The transport of gage damaged lattice density across surface $\{\underline{x}_s\}_G$ into the matrix will be neglected.
With this discussion, and using the total time rate of change expression of Eq(4), the evolution equation for damaged lattice density of the gage material is given by:

$$dD_G(\underline{x}, t; \underline{q})/dt = [\alpha_{p\ [G\text{-}Collision]}\ v_{p\ [Rado]}\ \mathbf{N}_{Rad}(\underline{x}, t; \underline{v}_{Rado})\ N_G(\underline{x},t; \underline{q}^*)$$

$$+ [\alpha_{p\ [M\text{-}Collision]}\ v_{p\ [Rado]}\ \mathbf{N}_{Rad}(\underline{x}^{M^*}, t; \underline{v}_{Rado})\ N_M(\underline{x}^{M^*}\ t; \underline{q}^*)\{\Delta X_{k\ DM\text{-}max}\}n_k/\{\Delta X_{h\ G}\}n_h \quad (5)$$

The approximation for damaged lattice density in the gage material of Eq(5) contains an explicit dependence on the finite length-scale for "knock-on transport" in the matrix material divided by the gage thickness length-scale, $\{\Delta X_{k\ DM\text{-}max}\}n_k/\{\Delta X_{h\ G}\}n_h$, and the "knock-on density of matrix scattering in the undamaged matrix lattice" is evaluated at a non-gage spatial point, $\underline{x}^{M^*}$, in the matrix material function, $N_M(\underline{x}^{M^*}, t; \underline{q})$. These coupling terms are not classical "continuum mechanics" response expressions and are finite length-scale effects of high-energy radiation scattering down path-lengths. The damage ratio, $\{\Delta X_{k\,DM\text{-}max}\}n_k/\{\Delta X_{h\,G}\}n_h$, of the matrix scattering length-scale to the gage length scale in the last term of Eq(5) means that the rate of damage lattice density is increased as the gage thickness decreases. Consequently, if radiation damage to the gage lattice density as related to "gage life-longevity performance" is a critical gage design requirement, then smaller and smaller nano-scale gage designs are intrinsically less and less robust. On the other hand, if radiation damage to the gage lattice density as related to "gage threshold radiation-flux identification performance" is a critical gage design requirement, then smaller and smaller nano-scale gage designs are intrinsically more and more sensitive[Patent Disclosure Filed $\rho\beta\sigma A$-Oct05].

Solutions to the Lattice Density Evolution Equations

As mentioned after Eq(3a), the matrix and the gage material have bulk evolution equations for their undamaged lattice densities, $N_M(\underline{x}, t; \underline{q})$ and $N_G(\underline{x}, t; \underline{q})$ respectively, described by Eq(3a). The damaged lattice, $D_M(\underline{x}, t; \underline{q})$, in the matrix material is described by the Eq(3b). And due to the analysis for damaged lattice transport across a surface[10], the gage's damaged lattice density evolution is described by Eq(5). In this case, solutions to the four reduced equations require that initial lattice density conditions be prescribed; say $N_M = N_{M0}$; $N_G = N_{G0}$: $D_M = 0$; and $D_G = 0$; at time t=0. But no prescribed displacement boundary conditions are required. With these conditions, one can write solutions, in the sense of a Liouville's species density moving in the phase space $\{\underline{q}(t)\}$, to Eqs(3) and Eq(5) for the density functions $N_M(\underline{x}, t; \underline{q})$, $N_G(\underline{x}, t; \underline{q})$, $D_M(\underline{x}, t; \underline{q})$, and $D_G(\underline{x}, t; \underline{q})$ as:

$$N_M(\underline{x}, t; \underline{q}) = N_{M0}(\underline{x},0,\underline{q}(t)) \left\{ \exp[-\int_0^t [\alpha_{p[M\text{-}Collision]}\, v_{p\,[Rado]}\, \mathbf{N}_{Rad}(\underline{x}, s, \underline{v}_{Rado})\, ds] \right\} \qquad (6a)$$

$$N_G(\underline{x}, t; \underline{q}) = N_{G0}(\underline{x},0,\underline{q}(t)) \left\{ \exp[-\int_0^t [\alpha_{p[G\text{-}Collision]}\, v_{p\,[Rado]}\, \mathbf{N}_{Rad}(\underline{x}, s, \underline{v}_{Rado})\, ds] \right\} \qquad (6b)$$

$$D_M(\underline{x}, t; \underline{q}) = N_{M0}(\underline{x},0,\underline{q}(t) +\Delta\underline{q}\,)\left\{ \int_0^t \left[[\alpha_{p\,[M\text{-}Collision]}\, v_{p\,[Rado]}\, \mathbf{N}_{Rad}(\underline{x}, r, \underline{v}_{Rado})] \right] \right.$$
$$\left. [\exp[-\int_0^r [\alpha_{p[M\text{-}Collision]}\, v_{p\,[Rado]}\, \mathbf{N}_{Rad}(\underline{x}, s, \underline{v}_{Rado})\, ds]]\, dr \right\} \qquad (6c)$$

$$D_G(\underline{x}, t; \underline{q}) = N_{G0}(\underline{x},0,\underline{q}(t)+\Delta\underline{q})\left\{ \int_0^t \left[[\alpha_{p\,[G\text{-}Collision]}\, v_{p\,[Rado]}\, \mathbf{N}_{Rad}(\underline{x}, r, \underline{v}_{Rado})] \right] \right.$$
$$\left[\exp[-\int_0^r [\alpha_{p[G\text{-}Collision]}\, v_{p\,[Rado]}\, \mathbf{N}_{Rad}(\underline{x}, s, \underline{v}_{Rado})\, ds]]\, dr \right\}$$
$$+ N_{M0}(\underline{x}^{M*},0,\underline{q}(t)+\Delta\underline{q})\left\{ \int_0^t \left[\alpha_{p\,[M\text{-}Collision]}\, v_{p\,[Rado]}\, \mathbf{N}_{Rad}(\underline{x}^{M*}, r, \underline{v}_{Rado})\, \{\Delta X_{k\,DM\text{-}max}\}n_k/\{\Delta X_{h\,G}\}n_h \right] \right.$$
$$\left. [\exp[-\int_0^r [\alpha_{p\,[G\text{-}Collision]}\, v_{p\,[Rado]}\, \mathbf{N}_{Rad}(\underline{x}^{M*}, s, \underline{v}_{Rado})\, ds]]\, dr \right\} \qquad (6d)$$

The embedded nano-scale gage solutions for the four density functions $N_M(\underline{x}, t; \underline{q})$, $N_G(\underline{x}, t; \underline{q})$, $D_M(\underline{x}, t; \underline{q})$, and $D_G(\underline{x}, t; \underline{q})$ provided in Eqs(6) are based on the simplifying approximations made to derive Eqs(3) and Eq(5). The solutions are all exponential integral responses in time that depend explicitly on the radiation flux field, $v_{p\,[Rado]}\, \mathbf{N}_{Rad}(\underline{x},t; \underline{v}_{Rado})$, the material scattering characteristics of the gage and matrix materials, $[\alpha_{p\,[G\text{-}Collision]}, \alpha_{p\,[M\text{-}Collision]}, \{\Delta X_{k\,DM\text{-}max}\}n_k]$, and the dimension of the embedded gage, $\{\Delta X_{h\,G}\}n_h$. These lattice density solutions will be used to describe the deformation response of matrix and gage materials in the next section.

Undamaged and Damaged Lattice Evolution Dependent Deformation Kinematics

A relative deformation functional that accounts for the "elastic $\Delta\underline{q} = \underline{q}_o+ \int_{\underline{\dot{q}}} dt$ and damage $\Delta\underline{q}$" changes in lattice size dimensions occurring in the undamaged and damaged lattice cells between two arbitrary spatial points A and B is derived from the relative position vector given at any arbitrary time t in Eq(2). This is done by first taking the time derivative of the relative position vector of Eq(2), and adding the lattice deformation rate contributions due to the evolution of the two lattice density functions to the integrand of relative velocity path integration:

$$\partial\{X_k(A{\Rightarrow}B, t: \{N(\underline{x}, t; \underline{q})\},\{D(\underline{x}, t; \underline{q})\})\}/\partial t = v_k(A{\Rightarrow}B, t: \{N(\underline{x}, t; \underline{q})\}, \{D(\underline{x}, t; \underline{q})\})$$

$$= \int_{\{A{\Rightarrow}B, t\}} \int_{\{\underline{q}\}}[\; a_k E_{pmn}b_m c_n + b_k E_{pmn}c_m a_n + c_k E_{pmn}a_m b_n\;]\; \partial\{N(\underline{x}, t; \underline{q}) + D(\underline{x}, t; \underline{q})\}/\partial t$$

$$+ [\partial a_k/\partial t E_{pmn}b_m c_n + \partial b_k/\partial t E_{pmn}c_m a_n + \partial c_k/\partial t E_{pmn}a_m b_n\;]\{N(\underline{x}, t; \underline{q}) + D(\underline{x}, t; \underline{q})\}d\underline{q}dx_p \qquad (7)$$

Eq(7) is a path integral functional for the relative velocity of material at arbitrary spatial point B with respect to arbitrary spatial point A in a matrix or gage material volume, $\{\underline{x}_R\}_M + \{\underline{x}_R\}_G$. Here, it is implicitly understood that if the spatial integration point $\underline{x}$ is an element in the matrix material volume $\{\underline{x}_R\}_M$ point set, then the functions in the integrand are elements from the matrix material function spaces, $\{N_M(\underline{x}, t; \underline{q})\}$ and $\{D_M(\underline{x}, t; \underline{q})\}$. Else the spatial integration point $\underline{x}$ is an element in the gage material volume $\{\underline{x}_R\}_G$ point set, then the functions in the integrand are elements from the matrix material function spaces, $\{N_G(\underline{x}, t; \underline{q})\}$ and $\{D_G(\underline{x}, t; \underline{q})\}$. And, in this latter case the damaged density function depends explicitly on the damaged lattice density of the matrix material transported into the gage material as described in the above analysis. Physically, the relative velocity value exists independent of the integration path, so for mathematical purposes the spatial path of integration is arbitrary. This arbitrariness of path integration will be useful in developing consistent stochastic metrics as lattice density dependent functional expressions to describe deformation kinematics, such as strain tensors and velocity gradients tensors. For example, the velocity gradient is, in classical continuum models, the spatial derivative of the local material velocity, but in a lattice cell model the spatial points are continuous; however, the material lattice is discrete at a spacing prescribed by the lattice species $\underline{q}$. So the classical "continuum functions" and its continuous spatial gradients are not mathematically consistent with the physical problem of discrete atomic lattice structures and discrete discontinuities in describing the kinematics of motion as the damage and undamaged lattice density functions evolve during radiation. The conceptual consistency between rigorous mathematically continuity and physically discrete events occurring in a spatial and/or temporal sequence is always an approximation. The approximation becomes better and better as the number density of discrete lattice events, measured per unit spatial volume per unit species volume per unit time interval dimensional length scales, becomes greater and greater.

The functional for displacement of position B relative to A is the integration over time of Eq(7):

$$u_k(A{\Rightarrow}B, t: \{N(\underline{x}, t; \underline{q})\}, \{D(\underline{x}, t; \underline{q})\}) = \int_{\{0 {\Rightarrow} t\}} v_k(A{\Rightarrow}B, s: \{N(\underline{x}, s; \underline{q})\}, \{D(\underline{x}, s; \underline{q})\})ds =$$

$$\int_{\{0 {\Rightarrow} t\}} \int_{\{A{\Rightarrow}B, s\}} \int_{\{\underline{q}\}}[\; a_k E_{pmn}b_m c_n + b_k E_{pmn}c_m a_n + c_k E_{pmn}a_m b_n\;]\;\{\partial N(\underline{x}, s; \underline{q})/\partial s + \partial D(\underline{x}, s; \underline{q})/\partial s\}$$

$$+[\partial a_k/\partial s E_{pmn}b_m c_n + \partial b_k/\partial s E_{pmn}c_m a_n + \partial c_k/\partial s E_{pmn}a_m b_n]\{N(\underline{x},s;\underline{q})+D(\underline{x},s;\underline{q})\}d\underline{q}dx_p ds \qquad (8)$$

For a small strain model, a stochastic undamaged-damaged dependent strain tensor γ_{kp} is derived directly from Eq(8) for the damaged matrix or the damaged gage material. The spatial derivatives of the relative displacement functional are defined by using the path independence of the spatial integration and the arbitrariness of point B relative to the arbitrary point A at spatial point $\underline{x}_A$. The mathematical existence of the spatial derivatives is formally assured for spatially integrate-able density function N and D; as the derivative operation is with respect to arbitrary path integration increments in spatial vector Δx_p. Thus, the small strain tensor functional is formally written as:

$$\gamma_{kp}(\underline{x}_A,\ t:\ \{N(\underline{x}_A,t;\ \underline{g})\},\ \{D(\underline{x}_A,t;\ \underline{g})\}) = \tfrac{1}{2}\{\ \partial_p u_k(\underline{x}_A,\ t:\ \{N\},\ \{D\}) + \partial_k u_p(\underline{x}_A,\ t:\ \{N\},\ \{D\})\ \} \quad (9)$$

For shorthand notational purposes, three integral functional operators A_{kp}, $\mathring{A}_{kp}$, and ΔA_{kp} are defined for lattice species integrations over the lattice species domain $\{\underline{g}\}$ as:

$$A_{kp}\{...\} = \int_{\{\underline{g}\}} [\ a_k E_{pmn} b_m c_n + b_k E_{pmn} c_m a_n + c_k E_{pmn} a_m b_n\ ]\ \{...\} d\underline{g} \quad (10a)$$

$$\mathring{A}_{kp}\{...\} = \int_{\{\underline{g}\}} [\partial a_k/\partial s E_{pmn} b_m c_n + \partial b_k/\partial s E_{pmn} c_m a_n + \partial c_k/\partial s E_{pmn} a_m b_n]\{...\}\ d\underline{g} \quad (10b)$$

$$\Delta A_{kp}\{...\} = \int_{\{\underline{g}\}} [\ \Delta a_k E_{pmn} b_m c_n + \Delta b_k E_{pmn} c_m a_n + \Delta c_k E_{pmn} a_m b_n\ ]\ \{...\}\ d\underline{g} \quad (10c)$$

Using the operators defined in Eqs(10), the gage material strain tensor can be compactly written as:

$$\gamma_{kp}(\underline{x},\ t:\ \{N_G(\underline{x},t;\ \underline{g})\},\ \{D_G(\underline{x},t;\ \underline{g})\}) = \tfrac{1}{2} \int_{\{0 \Rightarrow t\}} \{[\ A_{kp} + A_{pk}\ ]\ \{\ \partial N_G(\underline{x},\ s;\ \underline{g})/\partial s + \partial D_G(\underline{x},\ s;\ \underline{g})/\partial s\ \}$$
$$+ [\ \mathring{A}_{kp} + \mathring{A}_{pk}\ ]\ \{\ N_G(\underline{x},s;\underline{g}) + D_G(\underline{x},s;\underline{g})\ \}\ \}\ ds \quad \text{for } \underline{x} \text{ in } \{\underline{x}_R\}_G \quad (11a)$$

A similar equation for the matrix material strain tensor is:

$$\gamma_{kp}(\underline{x},\ t:\ \{N_M(\underline{x},t;\ \underline{g})\},\ \{D_M(\underline{x},t;\ \underline{g})\}) = \tfrac{1}{2} \int_{\{0 \Rightarrow t\}} \{[\ A_{kp} + A_{pk}\ ]\ \{\ \partial N_M(\underline{x},\ s;\ \underline{g})/\partial s + \partial D_M(\underline{x},\ s;\ \underline{g})/\partial s\ \}$$
$$+ [\ \mathring{A}_{kp} + \mathring{A}_{pk}\ ]\ \{\ N_M(\underline{x},s;\underline{g}) + D_M(\underline{x},s;\underline{g})\ \}\ \}\ ds \quad \text{for } \underline{x} \text{ in } \{\underline{x}_R\}_M \quad (11b)$$

The gage material strain tensor is expressed in Eq(11a) as a functional that depends on density function spaces, $\{\ N_G(\underline{x},t;\underline{g}) + D_G(\underline{x},t;\underline{g})\ \}$, but the gage damaged density function space is also explicitly dependent on damaged lattice density transported from the surrounding matrix material as expressed in the solution of Eq(6d) for gage damage density $D_G(\underline{x},t;\underline{g})$. This result is analytically significant in that the potentially complex coupling of "gage" and "matrix" material characteristics during gage response testing will be less difficult to comprehend; and particularly simple "bulk" material parameter correlations of gage response data from radiation testing can be perhaps greatly improved. The aspect of material failure, and subsequent gage performance, due to deformation "incompatibility" between the matrix and gage material can be only partially analyzed based on the two material strain expressions given in Eqs(11). The pre-failure condition of "relative displacement compatibility" is mathematically complex, but requires that the gage and matrix material be contiguously connected across the gage surface $\{\underline{x}_S\}_G$, up to the moment of failure. Thus, during pre-failure deformations, the gage and matrix material strain tensors are not required to be functionally continuous across the surface between the two distinct materials. However, the relative displacement functional across the gage surface, $\{\underline{x}_S\}_G$, is required to remain materially continuous; and gage failure occurs when the relative displacement functional first becomes materially discontinuous. The other surface condition that is required to be continuous is the traction vector that is typically expressed in terms of the "elastic stress tensor" and the surface unit normal vector. The evaluation of elastic stresses requires that the above gage and matrix strain tensors be decomposed into an "elastic" tensor and an "non-elastic tensor" for each material. These decomposition analyses remain to be completed, but are a necessary step to establish a stress failure and/or strain energy density failure criterion. So the two strain expressions are a necessary and a useful step to develop, and to understand, nano-gage failure mechanics models.

Summary One of the objectives of this paper was to develop a lattice density approach based on classical statistical mechanics concepts that can replace the classical concept of using the total number of displaced atoms as a metric to correlate data from radiation damage. The physical and stochastic complexities of radiation damage on deformation kinematics, as expressed in Eqs(7), (8), and (11), can not be mathematically described nor determined with a scalar metric as represented by the total number of radiation displaced atoms. Another objective was to address finite-length scale dimensional effects that exist, and in some cases may exceed the dimensions of nano-scale material components embedded in a surrounding matrix material. The discussion and approximate model for damaged lattice density transport across a matrix-gage material interface illustrates that a continuum mechanics approach is not adequate to represent the physical phenomena that occurs during radiation damage models at lattice density length-scales; and add-on approximations are necessary. Finally, the path independence of the undamaged and damaged lattice density deformation kinematics is a useful analytical result to evaluate and decouple relative deformations and strain tensors in multi-material volumes. The detailed applications to failure analyses of nano-scale gage materials embedded in a matrix material remain to be completed.

References

1. G.H. Kinchin and R.S. Pease: 1955, Reports on Progress in Physics, V-18, pages 3-33.

2. R.B. Stout: Jun2004, Actinide Alpha-Decay in Spent Nuclear Fuels, ρβσA Rpt 0010.

3. L. Boltzmann: 1895, Lectures on Gas Theory, trans.1964 S. Brush, Un of Cal Berkeley Press.

4. G.E.Shilov and B.L Guerevich, 1977, Integral Measure and Derivative, Dover Press, NY.

5. E. Schrödinger: 1950, Space-Time Structure, Cambridge University Press, UK.

6. R.B. Stout: 2005, Paper #1081, 10th ICEM, 4-8Sept05, Glasgow, Scotland.

7. N.F. Britton: 2003, Essential Mathematical Biology, Springer-Verlag London Limited, UK.

8. E. Parzen: 1962, Stochastic Processes, Holden-Day, Inc., San Francisco, CA.

9. H.S. Rosenbaum: 1975, Microstructures of Irradiated Materials, Academic Press, NY.

10. R.B. Stout and T.K. Stout: Oct2005, Nano-Gage Radiation Damage Transport, ρβσA Rpt 0016.

Mater. Res. Soc. Symp. Proc. Vol. 887 © 2006 Materials Research Society 　　　0887-Q07-05

Effects of ion-beam irradiation on the L1$_0$ phase transformation and their magnetic properties of FePt and PtMn films

Chih-Huang Lai[1], Sheng-Huang Huang[1], Cheng-Han Yang[1], C.C. Chiang[1], S. H. Liou[2], D. J. Sellmyer[2], M. L. Yan[2], L, Yuan[2] and T, Yokata[3]

[1]Department of Materials Science and Engineering, National Tsing Hua University, 300, HsinChu, Taiwan.

[2]Department of Physics and Astronomy, University of Nebraska, Lincoln, NE 68588

[3]Department of Environmental Engineering of Materials, Nagoya Institute of Technology, Nagoya, Japan.

ABSTRACT

In this paper, we illustrate how to modify the structure and magnetic properties of L1$_0$ FePt and PtMn films using ion-beam irradiation. Highly ordered L1$_0$ FePt and PtMn phases were achieved directly by using 2 MeV He-ion irradiation without conventional post-annealing. A high ion-beam current density ($\sim \mu A/cm^2$) was used to achieve direct beam heating on samples. This irradiation-induced heating process provides efficient microscopic energy transfer and creates excess point defects, which significantly enhances the diffusion and promotes the formation of the ordered L1$_0$ phase. In-plane coercivity of FePt films greater than 5700 Oe could be obtained after disordered FePt films were irradiated with a He-ion dose of 2.4×10^{16} ions/cm^2. The direct ordering of FePt took place by using ion-irradiation heating at a temperature as low as 230 ℃. In PtMn-based spin valves, an L1$_0$ PtMn phase, a large exchange field and a high giant magnetoresistance (GMR) ratio (11%) were simultaneously obtained by using He-ion irradiation. On the other hand, Ge-ion and O-ion irradiation completely destroyed the ferromagnetism of FePt and the GMR of PtMn-based spin valves, respectively.

INTRODUCTION

To develop a high-density magnetic recording system, sensitive reading elements and high-coercivity media are essential. Currently, the readers are composed of a spin-valve structure in which an antiferromagnetic (AFM) layer provides a strong exchange bias to pin the adjacent ferromagnetic (FM) layer. A typical material used for the AFM layer is PtMn. The applications of exchange anisotropy stimulated extensive theoretical and experimental studies [1-2]. It has been found when increasing areal density of media, the super-paramagnetic limit becomes a critical issue, especially when the grain size is reduced to below 5 nm. L1$_0$ FePt possesses high magnetocrystalline anisotropy ($K_u \sim 7 \times 10^7$ erg/cc) [3] and is one of most promising candidates for ultra-high-density media. Both as-deposited PtMn and FePt are disordered fcc phases, and post-annealing is needed to transform the disordered to ordered $L1_0$ phase. The typical annealing temperature for PtMn is around 250 to 280 ℃, and that for FePt is higher than 500 ℃.

When magnetic recording systems forge ahead to a small scale, it is important, therefore, to develop novel processing methods. In recent years, ion irradiation has drawn much attention as a potential tool to modify the magnetic properties without affecting the surface topography. For example, changing the anisotropy of (Co/Pt) multilayers [4], varying the GMR ratio of (Fe/Cr) multilayers [5] and enhancing the magnetic moment of CoPt alloys [6] were reported. In this work, we investigated the ion-irradiation effects on the magnetic thin films composed of $L1_0$ phases. Two topics are discussed: (A) the FePt films and (B) the spin-valve structure with PtMn layers.

Typically, as-deposited FePt films of disordered fcc phase are magnetically soft. Post-annealing at temperatures higher than 500℃ is needed to develop an ordered $L1_0$ phase and to achieve a large coercivity. He-ion irradiation was used to control the degree of chemical ordering of FePt films [7]. The long range ordering factor S of sputtered FePt (001) films can be improved by using post-growth irradiation. But only a film which already has a partially ordered structure (S=0.4) can develop an induced long-range ordered phase. Ion irradiation can introduce an appropriate amount of atomic species and desired energy into the films. The energetic incident ions may create defects or lattice distortion and thus may promote phase transformation. In part (A) of the results and discussion, we report that the ordering of FePt $L1_0$ phase can be directly induced by 2 MeV He-ion irradiation. We purposely adjusted the beam current to a $\mu A/cm^2$ scale to study the beam heating effect and the energy transfer in FePt films. In addition, the comparisons were made between rapid thermal annealing (RTA) and ion irradiation. We also propose a method of magnetic patterning for patterned media by using ion irradiation.

Ion irradiation with lithographic masks has been proposed as an alternative technique for making patterned magnetic nanostructures, and for defining the widths of tracks in giant magnetoresistance (GMR) sensors [8]. In the case of ferromagnetic/antiferromagnetic (FM/AFM) exchange-biased bilayers, the exchange bias is sensitive to the interface and the AFM anisotropic energy. Accordingly, ion irradiation has been used to change both the magnitude and direction of the exchange field [9]. The exchange field can be either enhanced by the creation of defects, acting as domain-wall pinning sites, or can be suppressed to zero by the intermixing. C-ion irradiation has been applied to modify the exchange field and GMR ratios of PtMn- and NiMn- based spin valves. C-ion irradiation increases the exchange field because of the formation of PtMn(C) and NiMn(C) phases whose c/a ratio considerably less than that of the parent phases [10-11]. However, the GMR ratio decreased because of the C-residuals at the CoFe/Cu interfaces. In part (B) of the results and discussion, we demonstrate that a high exchange field and GMR ratio of the PtMn-based spin valves can be achieved by using 2 MeV He-ion without post-annealing. Furthermore, we propose a method of magnetic patterning to define the dimension of spin-valve sensors by using ion irradiation.

EXPERIMENTAL PROCEDURE

In part (A), 50 nm FePt films were deposited on SiO_2/Si substrates at room temperature by using co-sputtering from elemental targets. The composition of the FePt films is $Fe_{0.5}Pt_{0.5}$, measured by Rutherford Backscattering Spectrometry (RBS). The He-ion irradiation was performed on as-deposited disordered FePt films at ambient temperature with an energy of 2 MeV by using a KN accelerator. The beam current was set between 1.25 $\mu A/cm^2$ and 6 $\mu A/cm^2$. The temperature on the sample surface during irradiation was in-situ monitored by an attached thermocouple. Computer simulation SRIM code [12] enabled us to predict the depth profile of He ions, the energy loss, and the defect creation in the FePt films. The hysteresis loops were measured by using a vibrating sample magnetometer (VSM). X-ray diffraction (XRD) θ-2θ scans using the Cu K_α line provided the structural identification. The ordering factor S was directly deduced from the c/a ratio [13].

In part (B), the structure of spin-valve samples is Si// NiFeCr 5/ NiFe 3/ CoFe 1.5/ Cu 2.6/ CoFe 2.2/ PtMn 20/ NiFeCr 5 (unit: nm). The bottom and top NiFeCr layers are employed to improve the (111) texture and prevent oxidation, respectively. The composition of PtMn is $Pt_{50}Mn_{50}$. The as-deposited PtMn is a disordered fcc phase. The PtMn-based spin valves are irradiated with He-ion at ambient temperature using a KN accelerator. The beam current is 1.08 $\mu A/cm^2$, with a corresponding temperature at the sample surface about 190 $^\circ$C, as measured using an *in situ* thermocouple, which was placed directly on the sample surface during irradiation. The He-ion dose was varied from 6.96×10^{15} to 1.74×10^{16} ions/cm^2. An external field of 900 Oe was applied during irradiation to ensure the unidirectional anisotropy of PtMn. The acceleration voltage was 2 MeV; the calculated projected range (Rp) is about 7.22 μm and the straggle range (ΔRp) is about 0.227 μm, according to the SRIM simulation program. Hence, the majority of He-ions were implanted into the silicon substrates. On the other hand, 42 keV O-ion irradiation was performed using a NEC 9SDH2 accelerator following the irradiation of the spin-valve samples with 2 MeV He-ions. The hysteresis loops were measured using a vibrating sample magnetometer (VSM). X-ray diffraction (XRD) θ-2θ scans using the Cu K_α-line elucidated the film structure. The GMR ratio was measured using a four-point probe at room temperature.

RESULTS AND DISCUSSION

Part (A). Ion-irradiation effects on FePt films

The hysteresis loop of the FePt film irradiated at the beam current of 1.25 $\mu A/cm^2$ with an ion dose of 2.4×10^{16} ions/cm^2 is shown in Fig. 1. The coercivity of the irradiated sample is as high as 5700 Oe. The temperature on the surface of the irradiated FePt film during irradiation was about 230 ℃ due to direct beam heating. As shown in Fig. 2, XRD scans reveal the evolution of the structural ordering in the irradiated FePt film. The (001) superlattice peak was observed after irradiation. Also, the (111) fundamental peak of the irradiated FePt film shifted toward the higher angle side, which indicates that the fcc disordered phase has been partially transformed into the $L1_0$ ordered phase. For comparison, we annealed an unirradiated sample using a conventional vacuum furnace at 230 ℃ for 10 h. The ordered peaks of FePt after conventional furnace annealing is unapparent, as shown in Fig. 2, leading to a low coercivity of 1200 Oe (as shown in Fig. 1).

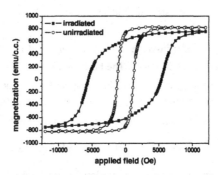

Figure 1. Hysteresis loops of irradiated and unirradiated FePt films. The He-irradiation was performed at a beam current of 1.25 $\mu A/cm^2$ with an ion dose of 2.4×10^{16} ions/cm^2. The unirradiated FePt film was annealed in a furnace at 230 ℃ for 10 h.

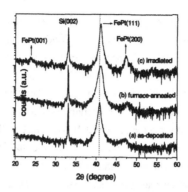

Figure 2. X-ray θ-2θ scans of (a) the as-deposited FePt film, (b) the unirradiated FePt film annealed in a vacuum furnace at 230 ℃ for 10 h, (c) the FePt film irradiated at a beam current of 1.25 μA/cm² with a dose of 2.4x10¹⁶ ions/cm². (The dash line is used to indicate peak shift).

The conventional post-annealing at 230 ℃ was not sufficient to provide enough energy for the ordering of FePt films. Instead, by using an ion-irradiation process, the energy transfer from He-ions to FePt films as well as point defects and lattice distortion induced by irradiation led to direct ordering. Conventional annealing only provides atoms extra thermal energy to overcome the activation energy of phase transformation. In contrast, the ion irradiation with a high beam-current-density not only provides beam heating but energy transfer from incident ions. The energy dissipation may cause the temperature to rise significantly within several nanometers around the moving ions and then rapidly quench to ambient temperature [14]. The atoms in the irradiated FePt films, therefore, have been relaxed from the original lattice points due to a succession of particle collisions and interaction. Combining the relaxation of the ion-disturbed FePt atoms by energy transfer with the beam heating to 230 ℃, the mobility of the FePt atoms has been enhanced considerably.

During the ordering process, the diffusivity of Fe and Pt atoms in FePt films determines the transformation rate and the ordering factor. According to SRIM code simulation, the defect concentration created in the irradiated FePt film by an ion dose of 2.4x10¹⁶ ions/cm² is 7.15x10²¹ defects/cm³ which is still much lower than the atom density of FePt. That is, FePt can still maintain its lattice without severe distortion. On the other hand, excess point defects such as vacancies and interstitials under irradiation are formed with a much higher concentration than that created by conventional annealing. The excess point defects can greatly increase the diffusivity of the Fe and Pt atoms. Consequently, irradiation-enhanced diffusion promotes the ordering of irradiated FePt films. Moreover, the sample surface is directly heated up by the ion beam, which further increases diffusivity. The irradiation-induced thermal heating process, providing a microscopic energy exchange and enhanced diffusion, is more efficient than any other conventional annealing process. As a result, a highly ordered $L1_0$ phase (S=0.73) of the FePt film was obtained with irradiation-beam heating at a low temperature of 230 ℃.

Fig. 3 illustrates the dependence of the coercivity H_c and ordering factor S on beam current for the samples irradiated at the same ion dose of 2.4×10^{16} ions/cm^2. By increasing beam current, the temperature on the sample surface was raised from 230 ℃ to 600℃. This is consistent with the W-ion irradiation in Si where the temperature of the ion flux is proportional to the beam current [15]. By increasing beam current, the beam heating effect is more pronounced and enhances the diffusivity so more volume fractions of ordered $L1_0$ phase were formed, leading to increased H_c and S.

Based on the previous results, we believe that the direct ordering of FePt can be obtained at the nominal temperature as low as 230 ℃ by using irradiation-heating process. This process involves several effects including microscopic energy transfer, excess point defects, and direct beam heating. We designed sets of experiments to understand which effects really affect the ordering of FePt. Porous anodic aluminum oxides (AAOs) were used as masks during irradiation. The AAO substrate with a thickness of 60 μm and a hole size of 100 nm on average is viewed as a beehive-like structure. 2 MeV He-ion irradiation with a beam current of 1.25 μA/cm^2 and an ion dose of 2.4×10^{16} ions/cm^2 was performed on the FePt/SiO$_2$/Si sample with a porous AAO

mask. The surface temperature of irradiated sample during irradiation was also 230°C.

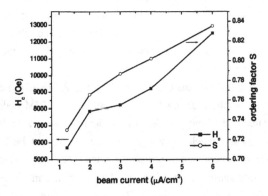

Figure 3. Dependence of coercivity H_c and ordering factor S on beam current for the samples irradiated at an ion dose of 2.4×10^{16} ions/cm^2.

The hysteresis loop of the irradiated sample with an AAO mask is identical to that of as-deposited samples. Since the aspect ratio (width to depth ratio) of the holes of porous AAO is quite large, only extremely straight and collimated incident He-ions can go through the porous AAO into FePt films. As a result, most He ions stopped in the AAO and the ordered FePt $L1_0$ phase cannot be obtained. In other words, although the ion beam directly heated up the sample, the microscopic energy transfer between incident ions and FePt atoms must be still required to achieve the ordering of FePt films. In addition, we performed Ni-ion irradiation with an energy of 8.8 MeV and a dose of 2.4×10^{16} ions/cm^2 on the FePt/SiO$_2$/Si films. The beam current was kept at 300 nA/cm^2. The temperature of the irradiated sample surface monitored by the thermocouple is 40 °C. Notice that all Ni ions penetrated through the film into the substrates according to the simulation of SRIM code. The energy loss in the FePt layer via electronic stopping was 800 eV/Å and the defect concentration was 2.25×10^{24} defects/cm^3, both of which are much higher than those of 2 MeV He-ion irradiation. The hysteresis loop of as-irradiated FePt films shows reduced H_c compared to as-deposited films. Although Ni-irradiation provides high energy transfer and excess point defects, FePt films still require direct beam heating to accelerate the diffusion and phase transformation. The reduced H_c could be due to reduction in domain wall energy by ion bombardment. Based on these results, all three effects: energy transfer, excess point defects and direct beam heating are essential for low-temperature ordering of $L1_0$ FePt by using ion irradiation.

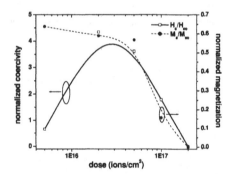

Figure 4. Dependence of normalized coercivity and magnetization on the Ge ion dose. Both values are normalized by the as-deposited disordered FePt values.

By using He-ion irradiation, a high coercivity of FePt can be obtained after irradiation. On the other hand, Ge-ion irradiation shows completely opposite effects. 75 keV Ge-ion irradiation was performed on the as-deposited disordered FePt films. The beam current was fixed at 300 nA/cm^2. The accelerating voltage was chosen to let a majority of Ge remain in the FePt films. The coercivity and saturation magnetization of the irradiated samples normalized by the values of as-deposited disordered films are shown in Fig. 4. When the dose was increased to 2×10^{17} ions/cm^2, the coercivity and magnetization approached zero. Consequently, Ge-ion irradiation destroyed the ferromagnetism of FePt. The irradiated Ge atoms may dilute the Fe concentration and reduce the Curie temperature to below room temperature.

Due to distinctively different results on the coercivity of FePt by using Ge and He ion irradiation, a novel approach of magnetic patterning is proposed to fabricate patterned media. The schematic diagram for this approach is illustrated in Fig. 5. A patterned FePt film was covered by resist, which has patterned metals on the top. The thickness and kind of metals are determined by the irradiation parameters and decide the depth profile of irradiated species in FePt films. The first step is to irradiate Ge-ions into the uncovered area (step (a)). The ferromagnetism of the irradiated area is destroyed by Ge-ion irradiation. The resist and metals are subsequently removed (step (b)). In step (c), He-ion irradiation is performed on the whole wafer. After irradiation, the wafer is divided into two regions, showing different magnetic signals during reading. The region, originally covered by metal and resist in step (a) during Ge-ion irradiation, shows a high coercivity and a strong magnetic signal because the L1$_0$ FePt phase was directly induced by He-ions in step (c). The other region, which is uncovered and exposed to Ge-ion irradiation in step (a), shows no magnetic signal. Consequently, the patterned media can be achieved by using ion irradiation without physical etching.

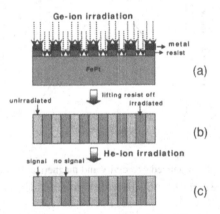

Figure 5. Schematic diagram of fabrication for patterned media by using ion irradiation. (a) Ge-ion irradiation is first performed, (b) resist and metals are removed and (c) He-ion irradiation is performed. The areas exposed to Ge-ion irradiation show no magnetic signal, but the rest

area shows a high coercity and magnetic signal.

Comparisons of rapid thermal annealing (RTA) and ion-irradiation were carried out on the [Fe(4.8 Å)/Pt(5 Å)]₁₀/C(100 Å) films. Typical in-plane and out-plane magnetization curves for the sample annealed at 550 °C for 300 secs by RTA method are shown in Fig 6 (a). The sample annealed at 550 °C by RTA has perpendicular anisotropy with a coercivity of 6 kOe. It also showed (001) texture in XRD spectra. The 2 MeV He-ion irradiated sample with a current-density of 5.04 $\mu A/cm^2$ for 360 secs, which corresponds to a surface temperature of 550 °C, has a coercivity as large as 10 kOe but no significant difference in the in-plane and out-of plane coercivity, as shown in Fig. 6 (b). All ion-irradiated samples have (111) textures, which lead to isotropic magnetic properties. These results may suggest that the structural evolution during the RTA and ion-irradiation process may be quite different, resulting in different textures. The structural modification by ion irradiation needs to be taken into account when ion irradiation is used for magnetic patterning.

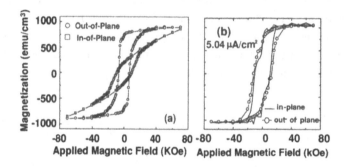

Figure 6. Magnetization curves for the sample (a) annealed at 550 °C for 300 sec by RTA method and (b) irradiated by using 2 MeV He-ions with current density of 5.04 $\mu A/cm^2$ for 360 secs

Part (B). Ion-irradiation effects on the spin-valve structure with PtMn layers

Figs. 7(a) and 8(a) show the hysteresis and MR loops, respectively, after He-irradiation. The irradiation dose is 1.91×10^{16} ions/cm². The exchange field and the MR ratio after irradiation are 650 Oe and 11%, respectively. For comparison, Figs. 7(b) and 8(b) present the hysteresis and MR loops of as-deposited samples, respectively. The exchange field and MR appeared after irradiation because of formation of the ordered PtMn phase, which was verified by the shift of the PtMn (111) peak in X-ray diffraction patterns. The MR ratio strongly depends on the interfacial conditions of CoFe/Cu/CoFe. The MR ratio of 11% is almost the same as that obtained by furnace annealing at 280 °C for 2 hours, which may demonstrate that He-ion irradiation does not induce severe interfacial changes because Rp is large. Similar to N-ion irradiation for current-in-plane (CIP) GMR pattering [8], He-ion irradiation can be used for defining sizes of CIP sensors composed of PtMn-based spin valves. The irradiated area shows a high GMR ratio, but non-irradiated area shows almost zero GMR. The schematic diagram of magnetic patterning for spin valve sensors is shown in Fig. 9 (a). Furthermore, the direction of unidirectional anisotropy can be controlled by changing the direction of the external field during irradiation. Consequently, ion irradiation combined with a lithographic mask can enable us to fabricate GMR cells with different pining directions on the same wafer. This process outperforms the conventional post-annealing process, which can only give single pinning direction on one wafer by an external field.

To eliminate MR and reduce the current shunting effect in the irradiated area, 42 keV O-ion irradiation was used to fabricate patterned GMR. First, GMR samples were irradiated with He ions at a fluence of 1.09×10^{16} ions/cm^2. An 11% MR ratio was achieved and O-ion irradiation was again applied. The MR ratio dropped and approached zero following O-ion irradiation at a fluence of 2×10^{16} ions/cm^2. Another approach for magnetic patterning for spin valve sensors is schematically shown in Fig. 9 (b).

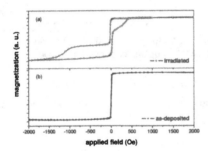

Figure 7. Hysteresis loops of (a) irradiated and (b) as-deposited spin valves. The irradiation fluence is 1.91×10^{16} ions/cm^2.

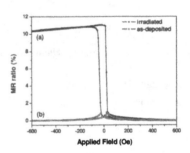

Figure 8. MR curves of (a) irradiated and (b) as-deposited spin valve samples. The irradiation fluence is 1.91×10^{16} ions/cm^2.

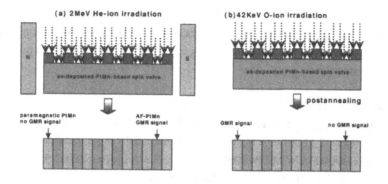

Figure 9. Schematic diagrams of magnetic patterning for PtMn-based spin valves by using (a) 2MeV He-ion irradiation and (b) 42 keV O-ion irradiation. The areas exposed to He ions in (a) show a high exchange field and GMR. The areas exposed to O-ion irradiation show no GMR.

We first irradiate O-ions into the uncovered area, then the whole wafer is annealed at temperatures higher than 260 °C to achieve ordered PtMn. The area exposed to O-ion irradiation still have no GMR signal after annealing, but the area without O-ion irradiation shows a high GMR. Furthermore, the resistance of the irradiated area significantly increased by a factor of approximately 100, indicating that O-ion irradiation oxidized the layer and eliminated GMR. Although the resistance of the irradiated area significantly increased, current shunting still existed. Other ion sources or higher O-ion fluence may be needed to further increase the resistance of irradiated area and to suppress current flow in the irradiated area. Combining conventional photolithography [8] or non-contact stencil masks [16] with ion irradiation, GMR films can exhibit a high MR ratio in the area with He-ion irradiation, and a zero MR ratio with a high resistance in the area with O-ion irradiation; therefore, the dimension of current-perpendicular-to-plane (CPP) GMR cells may be magnetically patterned by using ion irradiation.

CONCLUSION

We have demonstrated that the highly ordered FePt L1$_0$ phase can be obtained by using an irradiation-induced heating process. He-ion irradiation introduces (1) microscopic energy transfer between incident ions and FePt atoms, (2) excess point defects in FePt films and (3) direct beam heating. Combining these effects, the ordering can be achieved at a nominal surface temperature as low as 230 °C. In addition, we demonstrate that ion irradiation can be an alternative technique to fabricate patterned media.

We also showed that He-ion irradiation induced the direct-ordering of PtMn from fcc to the

$L1_0$ phase without post-annealing. An exchange field of 650 Oe and an MR ratio of 11 % were obtained after He-ion irradiation. O-ion irradiation can completely destroy GMR by effectively oxidizing layers. Combining He and O ion-irradiation provides magnetic patterning for PtMn-based spin valves.

ACKNOWLEDGMENT

This work was supported in part by the National Science Council of R.O.C. under Grant number NSC 93-2112-M-007-033, Ministry of Economic Affairs of R.O.C. under Grants 94-EC-17-A-08-S1-0006, by NSF MRSEC DMR-0213808, Army Research Office DAAD 19-03-1-0298, DOE, INSIC and the W. M. Keck Foundation.

REFERENCE

1. A. P. Malozemoff, Phys. Rev. B **35**, 3679 (1987).
2. D. Mauri, H. C. Siegmann, P. S. Bagus, and E. Kay, J. Appl. Phys. **62**, 3047 (1987).
3. K. Coffey, M. A. Parker, and J. K. Howard, IEEE Trans. Magn. **31**, 2737 (1995).
4. G. S. Chang, T. A. Callcott, G. P. Zhang, G. T. Woods, S. H. Kim, S. W. Shin, K. Jeong, C. N. Whang, and A. Moewes, Appl. Phys. Lett. **81**, 3016 (2002).
5. D. M. Kelly, I. K. Schuller, V. Korenivski, K. V. Rao, K. K. Larsen, J. Bottiger, E. M. Gyorgy, and R. B. van Dover, Phys. Rev. B **50**, 3481 (1994).
6. G. S. Chang, Y. P. Lee, J.Y. Rhee, J. Lee, K. Jeong, and C. N. Whang, Phys. Rev. Lett. **87**, 067208 (2001).
7. D. Ravelosona, C. Chappert, V. Mathet, and H. Bernas, Appl. Phys. Lett. **76**, 236 (2000).
8. L. Folks, J. E. E. Baglin, A. J. Kellock, M. J. Carey, B. D. Terris, and B. Gurney, IEEE Trans. Magn. **37**, 1730 (2001).
9. A. Mougin, T. Mewes, M. Jung, D. Engel, A. Ehresmann, H. Schmoranzer, J. Fassbender, and B. Hillebrands, Phys. Rev. B **63**, 060409(R) (2001).
10. C. H. Lai, C.H. Yang, R. T. Huang, C. W. Chen, F. R. Chen, J. J. Kai, H. Niu, J. Magn. Magn. Mater. **239**, 390 (2002)
11. C. H. Yang, C. H. Lai, S. Mao, J. Appl. Phys. **93**, 6596 (2003)
12. J. F. Ziegler, and J. P. Biersak, http://www.srim.org/.
13. B. E. Warren, X-Ray Diffraction (Addison-Wesley, Reading, MA, 1969) pp.208.
14. F. Seitz and J. S. Keohler, Solid State Phys. **2**, 305 (1956).
15. D. H. Zhu, H. B. Lu, K. Tao, and B. X. Liu, J. Phys.: Condens. Matter **5**, 5505 (1993).
16. B. D Terris, L. Folks, D. Weller, Appl. Phys. Lett. **75**, 403 (1999)

Degradation Processes at Nanometer Scale

Mater. Res. Soc. Symp. Proc. Vol. 887 © 2006 Materials Research Society   0887-Q11-01

Degradation in Sn Films Due to Whisker Formation

L. Reinbold[a], E. Chason[a], N. Jadhav[a], V. Kelly[a], P. Holmes[b], J.W. Shin[a], W.L. Chan[a], K.S. Kumar[a], G. Barr[b]
a) Brown University, Division of Engineering, Providence, RI 02912
b) EMC Corporation, Franklin, MA

ABSTRACT

Whisker formation in pure Sn coatings on Cu conductors is a serious impediment to the development of Pb-free electronics manufacturing. Understanding whisker formation is complicated by the fact that it is the result of multiple materials kinetic processes including interdiffusion, intermetallic formation and stress generation We report preliminary studies of whisker growth kinetics and stress evolution aimed at developing a fundamental understanding of the whisker growth process. A proposed model of point defect mediated stress generation provides a simple picture of how the different processes are connected.

INTRODUCTION

Thin films of pure Sn deposited on Cu can degrade by the formation of whiskers, i.e., thin filaments of Sn that can grow to lengths of 10's – 100's of microns. Commercially-used coatings on Cu conductors typically employ Pb-Sn alloys that do not develop whiskers. However, recent environmental regulations require the removal of Pb from electronics manufacturing . This has added a new urgency to understanding the creation of whiskers as the industry considers Pb-free alternatives to the current coating schemes.

The phenomenon of whisker formation has been known for many years [1], but the mechanisms and driving forces for whisker formation are still not fully understood. A large body of data indicates that many factors contribute to the formation of whiskers. Understanding their growth therefore requires understanding the interaction of multiple kinetic processes including interdiffusion, intermetallic formation and stress generation. To understand how these processes interact, we are studying the coevolution of stress, intermetallic formation, concentration profile and whisker growth on well characterized thin film samples.

In the current work, we report the results of preliminary experiments to study the driving forces and mechanisms controlling whisker formation. We present measurements of stress evolution (using wafer curvature) in vapor-deposited films of Sn on Cu and we also present measurements of whisker formation (using SEM) in films of Sn that have been partially covered with Cu. From these measurements (and others), we propose a picture of whisker formation in which the stress is mediated by the creation and diffusion of point defects.

BACKGROUND

Electroplated Sn and Sn-alloy finishes are used extensively in the electronic components industry to enhance solderability and corrosion resistance of electrical conductors made from metals such as copper and phosphor bronze. However, the growth of electrically conductive single crystal tin whiskers from the surface of pure tin coatings has been a technological problem for many years. These whiskers are capable of growing to lengths exceeding 10 mm and may carry 10-50 mA of current without burning off [2] – thus, they may easily bridge the gap between closely spaced interconnects and create a transient or permanent electrical short. The areal density of whiskers can be as high as 1000-10,000 units per square centimeter, and their induction period (i.e. time required for first visible indication of whisker formation) has been reported to vary from hours to years. Whiskers have been implicated in a number of system failures in military and commercial applications, including the loss of a Galaxy IV communications satellite [3,4].

There is still considerable disagreement about the mechanism of whisker formation. Multiple driving forces for whisker formation have been suggested, including recrystallization [5-9], oxide formation [10-16] and stress in the individual layers [10, 12-15, 17-26]. In models based on recrystallization, whisker growth is driven by surface free energy minimization. It is believed that this minimization is achieved by growing whiskers in specific directions and requires grain boundary immobility (so that whisker growth is the preferential means for decreasing free energy, rather than migration of the "pinned" tin grain boundaries). Formation of tin oxide on tin deposits has also been proposed to play a role in driving whisker growth. This was suggested as early as 1953 by F. C. Frank [11] and J. D. Eshelby [16], who assumed that the driving force for their dislocation mechanisms for whisker growth was surface oxidation that created a negative surface tension in the area from which a whisker eventually grew.

It is generally accepted that compressive stress in the film promotes whisker formation and growth. Stress may arise from different potential sources, including residual stresses that develop during the deposition process, diffusion of copper into tin, the formation of an intermetallic compound between the copper and tin, and even thermal expansion mismatch stresses that may develop under thermal cycling conditions. Various authors have set out to measure the internal stresses that develop in the Sn and Cu layers, using techniques ranging from flexure beam deflection to X-ray diffraction (XRD) and micro-diffractometry.

Lee and Lee [10] monitored the evolution of stress in electroplated tin deposits, and attributed the development of compressive stress in the tin to the diffusion of copper atoms into the tin matrix as well as the formation of Cu_6Sn_5 particles in the tin grain boundaries. The presence of an irregular (i.e. non-flat) interface between the Cu_6Sn_5 and tin layer was confirmed by stripping off the tin and imaging the sample cross-section via scanning electron microscopy (SEM). Zhang, et al. [24] also measured stress in tin deposits and confirmed Lee and Lee's observations that the Cu_6Sn_5 particles grow into the tin grain boundaries, asserting that the formation of this irregular interface is the key source of compressive stress that is directly responsible for driving whisker growth above the layer.

K.N. Tu [17] used XRD to study interdiffusion and intermetallic compound formation in vapor-deposited Sn-Cu thin film specimens. He attributed the stress generation and intermetallic compound formation to extremely rapid interstitial diffusion of Cu in Sn (and the inability of Sn to diffuse back into Cu). Tu [27] further suggested that long-range diffusion of tin atoms occurred to relieve the compressive stress in the tin film due to the formation of the intermetallic compound Cu_6Sn_5 at the interface between the Sn and the Cu substrate.

Choi et al. [12] utilized micro-diffractometry via synchrotron radiation to characterize stress states in the region around a whisker root. It was observed that the stress is highly inhomogeneous and varies from grain to grain (the stress is biaxial only when averaged over several grains). The stress gradient from the whisker root to the surrounding area was taken to indicate that the growth of the whisker relieves most of the local compressive stress. Recent work by Boettinger et al [26] used beam deflection to compare stress evolution and whisker formation on Sn-, Sn-Cu and Sn-Pb electrodeposited layers. This very thorough work included characterization of the different microstructures associated with the different layer compositions. Their results point to the importance of the layer composition in affecting the growth of the intermetallic particles and altering the stress state.

One difficulty with the stress-based models is understanding how the intermetallic growth induces the stress in the Sn layer. Lee and Lee [10] proposed that the compressive stress is due to the mechanical compression of the lattice by the expansion during growth of intermetallic particles at the grain boundaries. However, if the intermetallic forms primarily at the Cu-Sn interface, the stress generated by this mechanism would decrease rapidly away from the interface where the particles are forming. If this is the case, then it is not clear how the stress generated at the Cu-Sn interface would lead to formation of whiskers at the top surface of the sample, relatively far from this interface. Intermetallic formation may occur higher in the grain boundaries, but the majority of intermetallic generally forms near the Cu/Sn interface. In the discussion section, we propose a model in whish the stress is mediated by the long-range diffusion of point defects that addresses this issue.

EXPERIMENTAL PROCEDURE AND RESULTS

Bilayer samples of Sn and Cu were made by consecutive e-beam evaporation of the two metals in high vacuum at room temperature. Vapor deposition was used to enable precise control of the sample morphology and purity. The metal layers were deposited on oxidized Si substrates; a thin (100Å) Cr layer was deposited initially to enhance adhesion. Two types of sample film structures were grown, one consisting of uniform layers of Sn on Cu and one consisting of a partial Cu layer (ledge) over a continuous layer of Sn (this is shown schematically in Figure 1). The partially covered layer was created by depositing a 2000 nm layer of Sn uniformly on the Cr-covered substrate, and then using a shadow mask to cover half the Sn sample surface during subsequent deposition of 600 nm of Cu. The partially covered sample was used for SEM studies of whisker growth which, as discussed below, enabled us to compare whisker formation under Cu overlayers and in regions away from where intermetallic formation occurred. The uniform samples were used for monitoring the evolution of stress in the layers; typical Cu and Sn layer thicknesses were 600 nm for each layer.

The stress measurements were performed using an optical wafer curvature technique referred to as MOSS (multi-beam optical stress sensor). MOSS is a non-destructive technique that utilizes lasers to measure the curvature induced in a sample by stress in the thin film. This approach relates the curvature of the substrate to the product of the average stress and the layer thickness, σh (referred to as the stress-thickness). For multilayer structures, the stress-thickness can be summed over different layers to obtain the overall curvature. The MOSS technique is shown

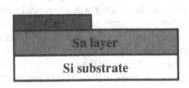

Figure 1. Schematic of Sn sample
partially coated with Cu overlayer.

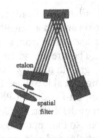

Figure 2. Schematic of MOSS (multi-beam
Optical Stress Sensor) for measuring stress.

schematically in figure 2. An array of parallel beams is reflected from the surface and deflection of the beams is measured by a CCD camera to monitor the curvature in real-time. The use of multiple beams significantly reduces the noise due to sample vibration, which makes it possible to measure stresses with resolution better than 1 MPa for the types of films used in this work. Because the technique is well suited for *in situ* measurements during processing, we have implemented MOSS systems on vacuum chambers for thin film evaporation and on electrochemical cells for electrodeposition or etching.

Stress measurements

Immediately following deposition, the film stress was monitored for a continuous four day period on the MOSS system (while the surface morphology and intermetallic compound formation were tracked periodically using SEM and XRD). Stress measurements for a Cu-Sn bilayer as well as pure Cu and pure Sn samples are shown in figure 3.

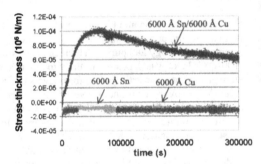

Figure 3. Real time stress measurements for pure Cu, pure Sn and 600 nm Sn/600 nm Cu bilayer.

The measured total curvature in the bimetallic layer initially corresponds to an overall tensile stress which becomes slightly less tensile (more compressive) over time. Note that the measured curvature in the bimetallic sample may be due to stress in either the Cu or Sn layers or both. In contrast, the pure Sn and Cu samples are each initially compressive and, after an initial relaxation, the stress stops changing after the first day.

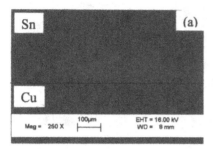 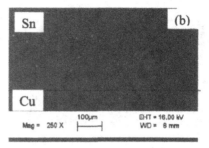

Figure 4. SEM micrograph of Sn layer partially covered by Cu overlayer (a) as-deposited and (b) after 38 days. The dotted line denotes the nominal boundary between the bare Sn surface and the Cu-covered region.

SEM of whisker formation in ledge sample

Scanning electron microscopy (SEM) was used to monitor the sample surface and whisker growth on the partially covered samples (referred to as "ledge samples"). The samples were measured as-deposited and at 24, 30 and 38 days after deposition. A comparison of the surface morphology of the as-deposited sample and the sample after 38 days is shown in figure 4. The dashed line is drawn onto the figure to indicate the nominal boundary between the Cu-covered region and the bare Sn. The whisker density is clearly higher after 38 days than in the as-deposited sample.

Whiskers were observed to form on both the Cu-covered portion of the sample and the part on which only Sn was deposited. Because of the relatively low number, it was difficult to accurately quantify the whisker density over the surface (further, more accurate studies are being conducted). However, the average whisker density was clearly higher on the bare Sn (approximately $110/mm^2$) than on the Cu-covered side (approximately $15/mm^2$). The density measured on the Cu side of the boundary appeared to increase between measurements made at 24 and 38 days but this may not be statistically significant. In comparison, the whisker density measured on the Sn side near the boundary did not change significantly between the measurements made at 24 and 38 days. The distance from the farthest observed whisker on the bare Sn side to the nominal Sn/Cu boundary was observed to increase with time.

The presence of whiskers on the Cu-covered side demonstrates that Sn whiskers are able to grow through the Cu overlayer. We have observed the growth of whiskers for layers of Cu as thick as $1\mu m$. A higher magnification image of a whisker growing through the Cu overlayer on the Cu-covered side of the sample is shown in figure 5a. EDS (energy dispersive spectroscopy) spot scans were performed along the body of the whisker, as well as the surrounding surface material. These measurements indicate that the cap on top of the whisker consists of Cu while the body of

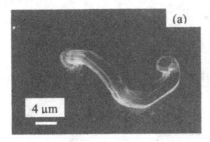 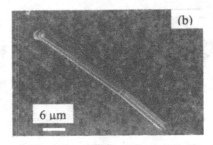

Figure 5. SEM images of whiskers observed on the (a) Cu-covered side of the sample and (b) on the bare-Sn side of the sample in figure 4b (38 days after deposition).

the whisker is composed of pure Sn. In order to grow, it appears that the Sn whisker pushes a chunk of Cu out of the overlayer like a hydraulic ram.

For comparison, a higher magnification image of a whisker growing on the bare Sn side of the sample is shown in figure 5b. On average, the whiskers on the bare Sn side have similar diameters to those growing on the Cu side (approximately 3μm.) but are slightly shorter (by approximately 30%). However, because of the higher density of whiskers on the bare Sn side we estimate that approximately three times as much Sn is extruded in the form of whiskers on the bare Sn side than on the Cu-covered side.

EDS was also used to monitor the Cu concentration profile across the sample surface over time. In the as-deposited sample, the boundary between Cu-covered and the bare Sn regions is not perfectly abrupt; the Cu deposition extends over the intended bare Sn region slightly due to the imperfection of the masking. The Cu concentration drops off to below measurable limits in an approximately linear way over a distance of approximately 100 μm from the nominal boundary. Over time, the Cu concentration profile on the surface was observed to extend further into the bare-Sn region (to at most 200 μm), presumably due to surface diffusion. However, it is important to note that whiskers were observed at a distance as far as 1 mm from the boundary where no Cu was detectable, significantly further than the region in which Cu was observed.

DISCUSSION

One conclusion that can be drawn from this work is the importance of the interaction between Cu and Sn for the formation of whiskers. Although it has been suggested that oxidation can be a driving force for whisker formation, we find that whiskers from under a Cu overlayer where they are presumably not oxidized. Other studies in our lab (not shown) have demonstrated that whiskers will also grow in vacuum of 1×10^{-9} torr where there is no significant oxidation.

Based on these results, we believe that the driving force for the whiskers is stress. Since we find that the stress in pure Sn and pure Cu layers is small, the as-deposited stress is not a significant source of stress. The stress therefore seems to be induced by the formation of the intermetallic in the Sn layer. Although this has been suggested by many others, what is not clear is how the volumetric difference between the intermetallic particle and the surrounding Sn matrix is converted into a stress that can lead to whisker formation. Lee and Lee [10] proppse that an irregular layer of Cu_6Sn_5 intermetallic compound beneath the Sn surface induces a mechanical stress on the surrounding Sn lattice which then drives whiskers to grow on the Sn surface. However, this explanation would suggest that the driving force should decrease with distance away from the intermetallic particles. This makes it difficult to understand how intermetallic formation at the interface between the Cu and Sn layers can create sufficient stress at the surface to form whiskers. This is made even clearer by the observations of whisker formation in the ledge sample. Whiskers form on the bare Sn surface far from the intermetallic, even in regions where no Cu is observed over the Sn.

Based on these observations, we propose that the stress is not created by elastic interactions but is instead due to the generation of point defects which can diffuse over long distances to create stress in regions far from where the intermetallic is forming. The proposed process is shown schematically in figure 6 for a layer of Cu deposited over the Sn. The point defects may be ejected from the intermetallic during its formation when Cu diffuses into the intermetallic and replaces the Sn. These defects may become interstitials, combine with Sn vacancies or diffuse along the grain boundaries. If they go to vertical grain boundaries or annihilate at dislocations, they can generate compressive stress in the layer that can be spread more uniformly throughout the layer vertically than just due to mechanical deformation at the interface. Note that if these point defects were able to annihilate at the free surface or horizontal grain boundaries, their creation would not lead to the generation of compressive stress in the film. Therefore, this suggests that the defects are precluded from annihilating at the free surface, perhaps by the presence of the surface oxide (or Cu overlayer, in the ledge experiments). In fact, in results not shown here, we have found that removal of the oxide by sputtering leads to relaxation of the stress in the Sn layer.

This model explains many of the features of whisker growth observed on the ledge samples. The stress builds up under the Cu layer first due to the generation of point defects in the Sn layer under the Cu. When the stress becomes sufficiently large, a Sn whisker is able to push its way through the Cu overlayer, emerging with a Cu ball on top of it. Over time, the point defects migrate into the bare Sn region and induce whisker formation there. The distance of the furthest

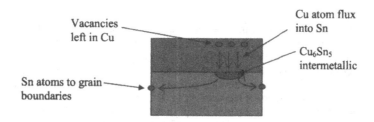

Figure 6. Schematic model of stress generation due to generation of Sn point defects.

whisker from the Cu/Sn boundary increases with time as the defects diffuse away into the Sn region. The density of whiskers on the Cu-side is lower than on the bare Sn-side due to the fact that it is more difficult for Sn whiskers to grow out through the Cu overlayer than through the Sn oxide.

This picture is also consistent with the stress measurements in the bilayer film. It has been well established that Cu diffuses rapidly into Sn but that Sn does not diffuse in the Cu layer. This leads to generation of tensile stress in the Cu layer as Cu diffuses into Sn and compressive stress in the Sn (and possibly intermetallic) layer as the intermetallic forms. It is the net sum of these stress-induced curvatures that is experimentally measured on the MOSS system. A schematic breakdown of these stress components is shown in figure 7, indicating how the observed curvature is the sum of different stress generation mechanisms. In order to know the stress in the Sn layer alone (which is the actual driving force for whisker growth), further experiments are necessary to separate the different components of the stress generation. For instance, the change in curvature when the Sn layer is removed by etching can be used to measure the average stress in the Sn layer.

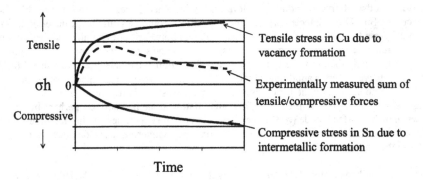

Figure 7. Schematic breakdown of various components contributing to resultant stress values measured by MOSS.

CONCLUSIONS

The results presented here are only preliminary because they require more systematic and comprehensive studies to quantify the effects we have observed. However, these results already suggest some important features of whisker growth. In particular, we think that the studies on the ledge sample confirm that the whiskers are the result of the Cu-Sn reaction to form intermetallic. The fact that whiskers develop far from the point where the intermetallic forms suggests that the stress generation is mediated by the long range diffusion of point defects. The evolution of stress in the bilayer samples is complex, but can be understood as a balance between tensile stress generation in the Cu (due to vacancy formation) and compressive stress generation in the Sn. Studies to measure the stress in the Sn directly by measuring the change in curvature when the Sn layer is chemically etched away are currently being conducted to verify this picture.

In the future, we will monitor the intermetallic formation simultaneously with the other parameters in order to quantitatively establish the relationship between stress, intermetallic volume and whisker growth kinetics. Comparing the evolution in layers deposited by other means (electrodeposition) and other compositions will enable us to probe the effect of microstructure and other kinetic barriers on the whisker formation.

ACKNOWLEDGEMENTS

The authors appreciate helpful discussions with Ben Freund, Allan Bower and Vivek Shenoy and gratefully acknowledge the support of Brown University's Materials Research Science and Engineering Center (DMR0079964) for this research.

REFERENCES

[1] H.L. Cobb, "Cadmium Whiskers", Monthly Rev. Am. Electroplaters Soc., 33 (28): pp. 28-30 (1946).

[2] J. Brusse, "Tin Whiskers: Revisiting an Old Problem", NASA's EEE Links Newsletter, 4 (4), pp. 5-7 (December 1998).

[3] S. Silverstein, "Reasons for Failure Lost with Galaxy 4", Space News, pp. 3, 20 (August 17-23, 1998).

[4] NASA Goddard Space Flight Center Tin Whisker Experimentation Page, www.nepp.nasa.gov/whisker/experiment/index.html

[5] V.K. Glazunova, "A Study of the Influence of Certain Factors on the Growth of Filamentary Tin Crystals", translated from Kristallografiya, 7(5): pp. 761-768 (1962).

[6] W.C. Ellis, D.F. Gibbons, R.C. Treuting, "Growth of Metal Whiskers From the Solid", Growth and Perfection of Crystals", ed. R.H. Doremus, B.W.Roberts, and D. Turnbull, New York: John Wiley & Sons, pp. 102-120 (1958).

[7] B.D. Dunn, "A Laboratory Study of Tin Whisker Growth", European Space Agency (ESA) Report STR-223, pp. 1-51 (1987).

[8] K. Tsuji, "Role of Grain Boundary Free Energy & Surface Free Energy for Tin Whisker Growth", Proc. of the IPC-Jedec Conf.-Frankfort, pp. 169-186 (2003).

[9] I. Boguslavsky and P. Bush, "Recrystallization Principles Applied to Whisker Growth in Tin", Proc. of the 2003 APEX Conf.-Anaheim, CA., pp. S12-4-1 to S12-4-10 (2003).

[10] B.Z. Lee and D.N. Lee, "Spontaneous Growth Mechanism of Tin Whiskers", *Acta Metallurgica*, 46(10): pp. 3701-3714 (1998).

[11] F.C. Frank, "On Tin Whiskers", *Philosophical Magazine*, Vol. 44, p. 851 (1953).

[12] W.J. Choi, T.Y. Lee, and K.N. Tu, "Structure and Kinetics of Sn Whisker Growth on Pb-free Solder Finish", *Proc. IEEE Elec. Comp. & Tech. Conf.*, pp. 628-633 (2002).

[13] C. Xu, Y. Zhang, C. Fan,, J. Abys, L. Hopkins, and F. Stevie, "Understanding Whisker Phenomenon : Driving Forces for the Whisker Formation", *Proc. IPC SMEMA APEX Conf.*, pp. S-06-2-1 to S 06-2-6 (2002).

[14] G.T.T. Sheng, C.F. Hu, W.J. Choi, K.N. Tu, Y.Y. Bong, and L. Nguyen, "Tin Whiskers Studied by Focused Ion Beam Imaging and Transmission Electron Microscopy", *J. of Appl. Phys.*, 92(1): pp. 64-69 (2002).

[15] C. Xu, Y. Zhang, C. Fan, and J. Abys, "Understanding Whisker Phenomenon: The Driving Force for Whisker Formation", Circuitree, pp. 94-105, April 2002. Available at www.circuitree.com.

[16] J.D. Eshelby, "A Tentative Theory of Metallic Whisker Growth", *Phys. Rev.*, 91: pp. 755-756 (1953).

[17] K.N. Tu, "Interdiffusion and Reaction in Bimetallic Cu-Sn Thin films", *Acta Metallurgica*, 21(4): pp. 347-354 (1973).

[18] R.M. Fisher, L.S. Darken, and K.G. Carroll, "Accelerated Growth of Tin Whiskers", *Acta Metallurgica*. 2(3): pp. 368-372 (May 1954).

[19] R.R Hasiguti, "A Tentative Explanation of the Accelerated Growth of Tin Whiskers", *Acta Metallurgica (letters to the editor)*, 3(2): pp.200-201 (1955).

[20] U. Lindborg, "Observations on the Growth of Whisker Crystals from Zinc Electroplate", *Metallurgical Transactions A*, 6A: pp. 1581-1586 (1975).

[21] M.E. Williams, C.E. Johnson, K.W. Moon, G.R. Stafford, C.A. Handwerker, and W.J. Boettinger, "Whisker Formation on Electroplated SnCu", *Proc. of AESF SUR/FIN Conf.*, pp. 31-39 (2002).

[22] K.N. Tu and K. Zeng, "Reliability Issues of Pb-free Solder Joints in Electronic Packaging Technology", *Proc. IEEE Elect. Comp. & Tech. Conf.*, pp. 1194-1199 (2002).

[23] J. Chang-Bing Lee, Y-L. Yao, F-Y. Chiang, P.J. Zheng, C.C. Liao, and Y.S. Chou, "Characterization Study of Lead-free SnCu Plated Packages", *Proc. IEEE Elect. Comp. & Tech. Conf.*, pp. 1238-1245 (2002).

[24] Y. Zhang, C. Xu, C. Fan, J. Abys, and A. Vysotskaya, "Understanding Whisker Phenomenon: Whisker Index and Tin/Copper, Tin/Nickel Interface", *Proc. IPC SMEMA APEX Conf.*, pp. S06-1-1 to S06-1-10 (2002).

[25] Y. Zhang, C. Fan, C. Xu, O. Khaselev, J.A. Abys, "Tin Whisker Growth – Substrate Effect Understanding CTE Mismatch and IMC Formation", *IPC Printed Circuits Expo, SMEMA Council APEX Designers Summit* (2004).

[26] W.J. Boettinger, C.E. Johnson, L.A. Bendersky, K.-W. Moon, M.E. Williams, G.R. Stafford, "Whisker and Hillock formation on Sn, Sn–Cu and Sn–Pb electrodeposits", *Acta Mat. (2005) in press.*

[27] K.N. Tu, "Irreversible Processes of Spontaneous Whisker Growth in Bimetallic Cu-Sn Thin Film Reactions", *Phys. Rev. B*, **49**, 2030 (1994).

Mater. Res. Soc. Symp. Proc. Vol. 887 © 2006 Materials Research Society   0887-Q05-01

Surface Segregation in Multicomponent Clusters

Peter A. Dowben[1], Ning Wu[1], Natalie Palina[2], H. Modrow[2], R. Müller[3], J. Hormes[4], Ya.B. Losovyj[1,4]

[1] Department of Physics and Astronomy and the Center for Materials Research and Analysis, University of Nebraska – Lincoln, Lincoln, NE 68588-0111, pdowben@unl.edu
[2] Physikalisches Institut der Universität Bonn, Nussallee 12, D-53115 Bonn, Germany
[3] Institut für Physikalische Hochtechnologie, Albert-Einstein-Straße 9, D-07745 Jena, Germany
[4] Center for Advanced Microstructures and Devices, Louisiana State University, 6980 Jefferson Hwy, Baton Rouge, LA 70806, ylosovyj@lsu.edu

ABSTRACT

Nanostructured materials are not immune from surface segregation, as can be shown for solid samples made from nanosized $BaFe_{12-2x}Co_xTi_xO_{19}$ barium ferrite particles and a variety of free clusters. Both theory and experiment provide ample demonstration that very limited dimensions of very small clusters does not necessarily impart stability against surface and grain boundary segregation. In fact, with the larger surface to volume ratio in small clusters and lower average atomic coordination, we anticipate that compositional instabilities in small clusters will readily occur.

INTRODUCTION

Surface segregation is a long standing problem in materials science with great technological significance. The phenomenon of surface segregation is the preferential enrichment of one component of a multi-component system at a boundary or interface. Atomic size and lattice strain, bond strengths, and even magnetic ordering influence the extent of segregation. Surface segregation indicates that the surface enthalpy is different from the bulk and occurs at finite temperatures (or in the materials growth process) when barriers to diffusion are overcome. The difference in the total free energy of the surface with respect to the bulk is a consequence of the surface truncation to vacuum and the resultant breaking of symmetry.

Besides surface segregation, the free energy difference drives other phenomena in order to minimize the total free energy [1], such as surface relaxations and surface reconstructions. All these phenomena are related to changes in crystalline order in the surface, resulting in the creation of different electronic and magnetic properties at the surface with respect to the bulk. These different surface electronic structure signatures can be sometimes exploited to study surface segregation.

It is well known that surface segregation plays a crucial role in affecting the surface and interface polarization ferromagnets [2]. This is considered to be very important to many spintronic applications as polarization is strongly influenced by composition [3-8]. Evidence that surface composition affects spin polarization of potentially high polarization materials abounds [3,4,6,9-13]. One of the more studied examples of reduced spin-polarization due to surface compositional instabilities of the NiMnSb half-Heusler alloy [2,3,14]. Magnetism plays a role in other ways: segregation can alter the magnetic ordering at the surface and thus also alter the surface free energy.

Clusters pose a more extreme example of systems with large surface to volume ratios, and are thus far more susceptible to surface segregation. With different stable faces at a cluster surface, the segregation will not be uniform across the cluster surface.

Unfortunately, the surface composition of an isolated cluster surface is difficult to characterize, and supported clusters are difficult to examine without careful consideration of a number of substrate contributions. These complexities aside, we can show that van der Waals bonded clusters [15,16] and even strongly ordered ferromagnetic alloy metal clusters will exhibit considerable surface segregation. As a result, for cluster assembled materials one cannot assume that because of the small cluster size there must exist compositional uniformity. The opposite is true: the smaller the cluster size in cluster assembled materials, the more likely the resulting material resembles ensemble of different compositional materials.

CHARACTERIZING SURFACE SEGREGATION

The surface energy depends upon crystal orientation, the extent of crystallinity, defect and grain boundary concentrations, the temperature, surface reconstructions, and surface lattice relaxations. Surface segregation, surface reconstructions and lattice defects will all result in changes in the total energy, but are limited by very different kinetics.

It is now fairly well established that equilibrium segregation can extend beyond the "topmost" or outer most atomic layer to a few atomic layers away from the surface (into the bulk material). The characterization of surface segregation must therefore be not only surface sensitive, but composition to a depth of the order of a few atomic layers becomes essential. At present, these characterization requirements are met to a large extent by analysis techniques based on electron and ion spectroscopies [17-18]. These are Auger electron spectroscopy (AES) [19], angle-resolved X-ray photoelectron spectroscopy or angle-resolved ESCA i.e. electron spectroscopy for chemical analysis (ARXPS) [1,6,9-14,19], ion scattering spectroscopy (ISS) [20] and secondary ion mass spectrometry (SIMS) [21]. We have tended to focus on ARXPS as the technique used for characterizing the surface composition, as the technique is not "destructive", and among the most established methods to accurately assess surface composition. The technique does require modification if applied to the study of clusters.

In ARXPS surface composition can be determined with considerable accuracy since the effective probing depth becomes shorter as the emission angle is increased with respect to the surface normal. The experimental core level intensities for any two components from a multi-component alloy are acquired at several emission angles θ, usually from $\theta = 0°$ (normal emission) to $60°$ (off-normal emission). Then, a linear background contribution is systematically subtracted from each raw spectrum. The peak intensities are further normalized by the corresponding differential cross-section for emission and by the analyzer transmission function [1,9-12,22]. The experimentally normalized intensity ratio for any two elements A and B is thus given:

$$R^{exp}(\theta) = \frac{I(A)/\sigma_A}{I(B)/\sigma_B} \frac{T(E_A)}{T(E_B)}$$

(1)

where $I(A)$ and $I(B)$ are the measured core level intensities for elements A and B, σ_A and σ_B are the cross sections, and $T(E_A)$ and $T(E_B)$ are the transmission functions of the electron energy analyzer for elements A and B as a function of the corresponding photoelectron kinetic energies E_A and E_B. The measured transmission functions must be determined for each analyzer. (For example, the transmission function of PHI 10-360 Precision Energy Analyzer is $T(E_A) = (E_A)^{1/2}$ [23,24]. For the Gamma Scienta SES-100 the corresponding transmission functions have been determined to a little more complex [22]).

We can apply a simplified model to fit the experimental intensity ratios obtained from ARXPS. The comparison between theory and experiment is accomplished through considerations of the theoretical normalized intensity ratio of element A and B as given by:

$$R^{theor}(\theta) = \frac{\sum_{j=0}^{\infty} f_j(A) \exp[\frac{-jd}{\lambda_A^j \cos(\theta)}]}{\sum_{j=0}^{\infty} f_j(B) \exp[\frac{-jd}{\lambda_B^j \cos(\theta)}]}$$

(2)

where λ_A^j and λ_B^j are the inelastic mean free paths of the core electrons generated from elements A and B respectively and passing through the material contained in layer j. The inelastic mean free paths can be adopted from previously published methodologies [25-26]. The atomic fraction of element A (chosen as the element which segregates to the surface) in the jth layer below the surface is roughly given by:

$$f_j(A) = b + \delta \exp(-jd/G)$$

(3)

where b is the bulk fraction of element A, δ and G are fitting parameters representing the extent of the segregation and the segregation depth respectively, and d is the distance between atomic layers. These two quantities are also the fitting parameters when comparing the model with experimental values. From the profile form $f_j(A)$ one can calculate the apparent surface concentration (or relative intensity) of element A for a particular core level. For strongly ordered alloys, different profile forms must be employed, and consideration of the likely short range order parameters cannot be neglected. Phase separation near the surface can further complicate analysis.

These complications are highlighted in some multicomponent oxides, in particular the manganese perovskites [10-13]. In the case of $La_{0.65}Pb_{0.35}MnO_3$, with a gentle annealing procedure (up to 250 °C), the surface is dominated by appreciable Pb segregation, whereas a heavily annealed surface (up to 520 °C) undergoes an irreversible restructuring into a Ruddlesden-Popper phase $(La_{1-x}Pb_x)_2MnO_4$ [13], as indicated in Figure 1. For $La_{1-x}Ca_xMnO_3$ (x=0.1 and 0.35), the terminal layer is predominantly Mn-O for x=0.35, while for x=0.1, the majority of the surface is La/Ca-O terminated according to analysis of surface composition with XPS [10]. Sr surface segregation has also been found in $La_{0.65}Sr_{0.35}MnO_3$, which causes a major restructuring of the surface region characterized by the formation of a Ruddlesden-Popper phase $(La,Sr)_{n+1}Mn_nO3_{n+1}$ with n=1 [11,12].

For clusters, these techniques have to be modified. Angle resolved techniques have to be replaced by comparing different core levels, with different photoelectron kinetic energies are different effective mean free paths, or by comparing select photoemission states at different photon energies. The equations above can be modified to extract a composition profile, as has been done for segregation studies of Fe-Cr alloys [27-28], utilizing the different mean free paths of the signature photoelectron signals. At low kinetic energies, the mean free path should strongly scale as the inverse square of the free-electron plasmon energy ($\propto E_p^{-2}$) or, alternatively, roughly proportional to the inverse of the number of valence electrons [29-31] but only if we disregard the spin dependent plasmon density and the metallicity (insulators typically have far longer mean free paths for electrons than metals). The spin-polarized electron mean free path, however, has been ascribed to the number d band holes [32] and electron-electron scattering [33].

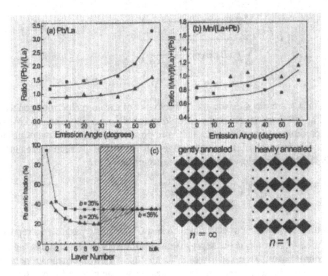

Figure 1. Panel (a) shows the normalized relative XPS core level intensity ratios of Pb($4f$) peaks to that of La($3d$) peaks and (b) shows the corresponding experimental and calculated intensity ratios of Mn to (La+Pb) core level intensities. The best fits (—) for the gently annealed surface (■) and for the heavily annealed surface (▲) of $La_{0.65}Pb_{0.35}MnO_3$, as a function of photoelectron emission angle (θ), are compared with the data. Panel (c) displays the resulting calculated Pb atomic fractions as a function of layer number. The inset shows schematically the layers stacking sequence used to obtain the compositional fits in panels (a) and (b) for the gently (left) and heavily (right) annealed surfaces. Adapted from [13].

There is a major problem that the usual means for estimating electron mean free paths in solids are not readily applicable to free and isolated clusters, particularly small clusters. As an example of the complexities of estimating electron mean free path in clusters, Auger electron emission from argon clusters are seen to be more surface sensitive than photoelectrons of a similar kinetic energy [34] and this cannot be easily related the number of valence electrons nor the plasmons.

There are examples of segregation in small free clusters, as described below, in spite of the difficulties in characterization surface composition of a very small free cluster, or cluster assembled material. Some studies of surface segregation, in clusters, have tended exploited the small difference in binding energy for atoms place at the surface compared to the bulk (the surface to bulk core level shift [35]). For cluster assembled materials, such a signature must consider the boundary between clusters as well as the surface. So this latter technique must be applied with care.

THE ENERGETICS OF SEGREGATION

Over the past many decades, the phenomenon of surface segregation attracted considerable theoretical interest. Various theories have been developed in order to account for the enrichment at the binary alloy surface. First-principles approach [36], embedded-atom method (EAM) model [37], Finnis–Sinclair (FS) potential [38], and Bozzolo–Ferrante–Smith (BFS) method [39] have all been used to simulate the surface segregation for fcc type random and ordered alloys, while for fcc and hcp lattices, the modified analytic embedded-atom method (MAEAM) many-body potential is used to theoretically study the surface segregation phenomena of binary alloys [40]. Unfortunately, there are no detailed finite temperature theoretical studies of the surface segregation of multi-component half-metal alloys, and as yet, little has been done to model non-equilibrium segregation in multi-component systems. Monte Carlo simulations have provided some insight and have more extensively employed to model segregation in clusters. It is a challenging vista of opportunity for both the theorist and experimentalist.

During the early 1980's on binary alloys, Moran-Lopez [41] found that they could predict between two metals, which would act as the solvent and which as a solute. Of the many binary systems they considered, it was only those which contained transition metal elements, which were known to order magnetically (ferromagnetic, anti-ferromagnetic, etc.) that deviated from expectation. Magnetic ordering, not often well understood at the surface of a cluster, must therefore be considered as well. Ternary systems containing components with magnetic moments are even more complex. For example, for two very similar ternary semi-Heusler alloys, NiMnSb and TiCoSb, the resultant segregation could not be predicted from classical mechanism arguments [14,42]. In the case of NiMnSb, it was found that Mn enriched the surface layers whereas for TiCoSb, Co and Sb dominated the surface and sub-surface regions, respectively.

In considering the influence of magnetic moments upon segregation, one must go back to basic ideas to identify what magnetic ordering would occur and justify why such ordering is energetically favorable. For Cr and Mn alone in bulk or thin layers, their spin coupling leads to an anti-ferromagnetic alignment; although, depending on crystal orientation, uncompensated spins can exist at the surface or buried interface. However, if Cr or Mn is alloyed with another element, ferromagnetic ordering is very possible, for example in $BiMnO_3$ and CrO_2. There is also the need to consider the effective Curie temperature, as with decreases in the ferromagnetic layer thickness there should be decreases in T_C (finite size scaling). These subtle possible contributions to the surface energy due to magnetic ordering are illustrated schematically in Figure 2.

As we have already noted, in order to minimize the free energy difference, surface segregation, surface reconstruction or/and surface relaxation compete, making most efforts to estimate the energetics of segregation in multicomponent systems, with more than two components, quite complex. Given that the presence of surface segregation is an indicator of a difference in free energy (chemical potential) between the surface and the

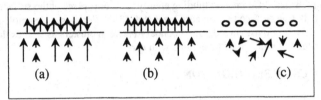

Figure 2. Surface ordering of moments in (a) anti-ferromagnetic ordering (b) ferromagnetic ordering (c) non-magnetic insulator models which could provide lowering of difference in surface free energy. Adapted from [42].

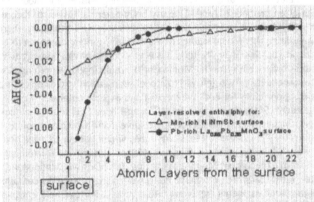

Figure 3. The enthalpy difference (ΔH) between the surface and the bulk based on segregations studies of NiMnSb [9,14] and $La_{0.65}Pb_{0.35}MnO_3$ [13].

bulk [1], estimates of the differences in the surface enthalpy relative to the bulk, based upon experiment can be quite valuable [2]. This energy difference is typically calculated from the experimentally determined segregation profiles using the approximation:

$$f_j(A)/f_j(B)= [f_b(A)/f_b(B)]exp(-\Delta H/k_BT) \qquad (4)$$

where $f_b(A)$ and $f_b(B)$ are the bulk layer concentrations of constituents A and B, and T is the annealing temperature [43]. These enthalpy differences exists between the surface and the bulk have been estimated for the half Heusler alloy NiMnSb(100) surface [9,14,42], based on most the extensive Mn segregated surface of NiMnSb, as well as for the Pb segregated surface of $La_{0.65}Pb_{0.35}MnO_3$ [13], as indicated in Figure 3, among other systems studied [1,2,10].

Any effort to extend this type of analysis to clusters is simplified by the small volumes, but when the cluster is too large to be amenable to a molecular calculation, simplifications are typically applied: shallow 4f or 5d levels are treated as core levels, and neglect of spin-orbit coupling, magnetic ordering energies, correlation energies, lattice relations are among the simplifications commonly undertaken. All of this tends to make us suspect that segregation in clusters is often more rapid and more extensive than one gets from an inspection of much of the cluster assembled materials literature.

SEGREGATION IN WEAKLY COHESIVE CLUSTERS

It is not necessary for there to be a strong chemical potential for surface segregation to occur in a cluster. A good example is the van der Waals "like" clusters. Radial surface segregation has been demonstrated in the self-assembled Ar/Xe and Ar/Kr clusters formed in an adiabatic expansion [15,16]. In both types of free gas clusters, there is a clear preference for the lower cohesive energy atoms to occupy the surface and leave higher cohesive energy atoms in the bulk. As a first approximation, the surface energy is proportional to the faction of missing bonds of the surface atoms and the cohesive energy. Consequently, quasi-spherical shapes and placing lower cohesive energy on the surface can minimize the surface cost.

Because of the much higher condensation temperature for Xe compared to Ar, little Xe is needed in the primary gas mixture to make the resulting free clusters, from gas expansion, mostly Xe. When the clusters are composed of both Xe and Ar atoms (or Kr and Ar atoms), the Ar 2p spectra tends to have a dominant signature of argon occupying surface sites. In Figure 4, the XPS spectra of the Ar 2p and Xe 4d core levels show that argon segregates to surface the Ar/Xe clusters from the proportionally greater surface contribution to the 2p core level [16]. For the higher Xe/Ar ratio (5.3% primary gas mixing contribution) dominant peaks in Ar 2p spectra I_{Ar} and S_{Ar} indicate that both Ar and Xe atoms both occupies the surface, corresponding to the high surface peak in Xe spectra. For a primary gas mixing ratio of 3.2% Xe, besides Ar-Xe mixed surface, increasing intensity of I_{Xe} and decreasing intensity of the surface Xe contribution suggests that a Xe interface layer is formed between the Xe bulk and Ar surface. With the increase of Ar, the mixture of Ar and Xe at the surface is replaced by almost all Ar [16].

Mixed Ar/Kr clusters exhibit similar radial surface segregation of Ar atoms at the surface while Kr atoms tend to dominate the bulk [15]. No interface layer has been observed in these mixed Kr and Ar clusters due to the relatively small difference of cohesive energy between Ar and Kr [15]. Resonant excitation processes, such as occurs in resonant photoemission, and near edge X-ray absorption fine structure spectroscopy, can significantly increase surface sensitivity. The shifts between the surface and bulk contributions, even in pure argon clusters are readily identified [44].

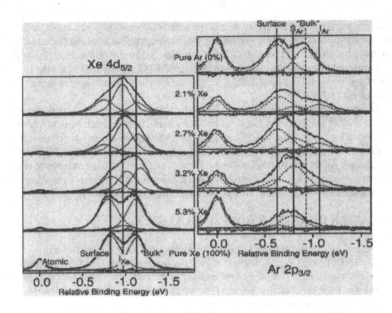

Figure 4. The XPS spectra of the Ar $2p_{3/2}$ and Xe $4d_{5/2}$ core levels for clusters produced from different primary gas mixing ratios. The peak at zero relative binding energy corresponds to the atomic peak. Adapted from [16].

SEGREGATION IN CLUSTERS WITH SHORT RANGE ORDER

Surface segregation is a known phenomenon in mixed component clusters, which can be treated as solid solution, but short range order can be extremely persistent in many alloy materials, causing notable deviations from an ideal solution. So is the strong ordering seen in bulk alloys observed at the surfaces of small isolated clusters of the same composition ? Generally this is not likely to occur.

FePt nanoparticles are considered as a promising candidate material for ultrahigh density magnetic storage media, because of the large magnetic anisotropy energies. Monte Carlo simulations have been applied on this binary metallic cluster system in the $L1_0$ phase [45,46]. Significant segregation is indicated if the difference of surface energy between pure elements Fe and Pt is considered [45]. This energy difference results in an energetic driving force for surface segregation. Such segregation will alter with temperature and cluster size, but can alter an ordered cluster from one with large nearest neighbor short range order to a cluster with a strong preference for surface enrichment by Pt (white) atoms [45], as indicated in Figure 5. A reduction of nanoparticles size can decrease ordering temperature, so that at as the cluster size decreases, disorder increases for the same finite temperature [46]. Because of the enhancement of surface to volume ratio with decreasing particle size, it is again preferential to place low surface energy atoms in the surface.

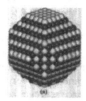 

Figure 5. A comparison of (a) an ordered 3.3 nm FePt nanoparticle, with (b) an equilibrium cluster above the ordering temperature. White is Pt and dark is Fe. Adapted from [45].

SEGREGATION IN $BaFe_{12-2x}Co_xTi_xO_{19}$ CLUSTER ASSEMBLED MATERIALS

Surface segregation can also be found in $BaFe_{12-2x}Co_xTi_xO_{19}$ barium ferrite nanoparticles [47]. Resonant photoemission results show that cobalt and titanium dopants strongly hybridize with the barium ferrite matrix. Due to the strong resonant behavior, we can find different element contributions to the various occupied valence bands in the system. It was found that the enhancement of resonance in 1.8 eV valence band can result from both barium and cobalt contributions. Extensive annealing (at 685 K for 10 hours in vacuum) of the sample leads to enhancement of the feature at 1.8 eV binding energy, indicating that barium and cobalt are enriched near the surface region, while iron is depleted at the surface compared to the bulk $BaFe_{12-2x}Co_xTi_xO_{19}$. This segregation is evident from the shift in the spectral peaks near 60 eV from the broad feature at 61 eV (unoxidized Co 3p is nominally at 58.9 eV binding energy, while unoxidized Ti 3s is nominally at 58.7 eV) to features at 58 eV and 56 eV binding energy (unoxidized Fe 3p is nominally at 52.7 eV binding energy). Figure 6 shows some of the evidence of the surface segregation phenomenon in this system. Such Co oxide segregation, following extensive annealing, does not require a magnetically dead layer at the surface of the crystallites as ferromagnetic Co doped titanium oxide is known.

While there has been some speculation concerning a dead layer in cobalt and titanium doped barium ferrites, from magnetic modeling [48], such a layer would be less than 1nm to be consistent with the photoemission results. The limited mean free path of the photoelectrons indicates that the segregation layer is quite thin, even after extensive annealing, consistent with the very short argon ion sputtering time needed to remove the segregation layer. Even if the segregation leads to a material that is nominally paramagnetic, solutions to the Landau-Ginzburg equation [49] indicate that a layer so thin (< 1 nm) would have an induced magnetization by proximity.

In this system the very large number of components, with competing contributions to the electronic structure, lattice strain, and the availability of a large number of different oxidation states, so magnetic ordering may not even play a dominant role in the surface segregation, but it cannot be neglected *a priori*.

SUMMARY

Surface segregation in alloy free clusters as well as surface and grain boundary segregation in cluster assembled materials must be taken as a given, unless there is sufficient evidence to indicate otherwise. Strong alloy ordering does not necessarily impart compositional stability to the surface, nor does weak inter-atomic interactions. Even at very low finite temperatures segregation may still occur, particularly if there is a high density of lattice vibrations, so segregation can occur even for "refractory" alloy clusters. Rigid refractory solids are not immune from surface segregation, with surface segregation evident at temperatures that are sometimes just a very small fraction of the melting temperature (2-3%). The tendency for surface segregation to occur in clusters may be exacerbated.

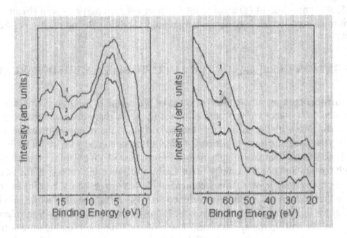

Figure 6. Photoemission spectra taken at room temperature after prolonged heating followed after several short heating/sputtering cycles: (1) prolonged annealing at 685 K, (2) after brief heating, (3) after sputtering. Spectra of the Co 3p and Fe 3p shallow cores (right panel), shows enhancement of Co 3p and/or Co and Ba Auger with annealing while the enhancement of the valence band 1.8 eV binding energy feature is shown at left. The photon energy was 115 eV. Adapted from [47].

ACKNOWLEDGEMENTS

The support of the Center for Advanced Microstructures and Devices (CAMD) and the Louisiana Board of Regents, the NSF "QSPINS" MRSEC (DMR 0213808), and the financial support by the DFG within the SPP1104 (contract no. Mo940/1) are gratefully acknowledged. We also acknowledged useful conversations with Olle Bjørnehom, Eckart Rühl and the contributions of C. Borca, A.N. Caruso and H.J. Jeong to the work.

References

[1] P.A. Dowben and A. Miller, editors, *Surface Segregation Phenomena*, (CRC Press, Boston, 1990) p. 145.

[2] P.A. Dowben and S.J. Jenkins, in *Frontiers in Magnetic Materials*, edited by Anant Narlikar (Springer Verlag, 2005), pp. 295-325.

[3] D. Ristoiu, J. P. Nozières, C. N. Borca, T. Komesu, H. –K. Jeong and P. A. Dowben, Europhys. Lett. **49**, 624 (2000).

[4] G. L. Bona, F. Meier and M. Taborelli, Solid State Commun. **56**, 391 (1985).

[5] W. Zhu, B. Sinkovic, E. Vescovo, C. Tanaka and J. S. Moodera, Phys. Rev. B **64**, 060403 (2001).

[6] C.N. Borca, B. Xu, T. Komesu, H.-K. Jeong, M.T. Liu, S.-H. Liou, S. Stadler, Y. Idzerda and P. A. Dowben, Europhys. Lett. **56**, 722 (2001).

[7] L. Ritchie, G. Xiao, Y. Ji, T. Y. Chen, C. L. Chien, M. Zhang, J. Chen, Z. Liu, G. Wu and X.X. Zhang, Phys. Rev. B **68**, 104430 (2003).

[8] A. Plecenik, K. Fröhlich, J. P. Espinós, J. P. Holgado, A. Halabica, M. Pripko and A. Gilabert, Appl. Phys. Lett. **81**, 859 (2002).

[9] D. Ristoiu, J. P. Nozières, C. N. Broca, B. Borca and P.A. Dowben, Appl. Phys. Lett. **76**, 2349 (2000).

[10] J. Choi, J. Zhang, S.-H. Liou, P.A. Dowben and E.W. Plummer, Phys. Rev. B **59**, 13453 (1999).

[11] H. Dulli, P.A. Dowben, S.-H. Liou, E.W. Plummer, Phys. Rev. B **62**, R14629 (2000).

[12] H. Dulli, E.W. Plummer, P.A. Dowben, J. Choi and S.-H. Liou, Appl. Phys. Lett. **77**, 570-572 (2000).

[13] C.N. Borca, Bo Xu, T. Komesu, H.-K. Jeong, M.T. Liu, S.-H. Liou, P.A. Dowben, Surface Science Letters **512**, L346 (2002).

[14] A.N. Caruso, C.N. Borca, D. Ristoiu, J.P. Nozières, P.A. Dowben, Surf. Sci. **525**, L109-L115 (2003).

[15] M. Tchaplyguine, M. Lundwall, M. Gisselbrecht, G. Öhrwall, R. Feifel, S. Sorensen, S. Svensson, N. Mårtensson, O. Björneholm, Phys. Rev. A **69**, 031201 (2004).

[16] M. Lundwall, M. Tchaplyguine, G. Öhrwall, R. Feifel, A. Lindblad, A. Lindgren, S. Sorensen, S. Svensson, O. Björneholm, Chem. Phys. Lett. **392**, 433-438 (2004).

[17] H.W. Werner and R P H Garten, Rep. Prog. Phys. **47**, 221 (1984).

[18] S. Hofmann, Surf. Interface Anal. **9**, 3 (1986).

[19] D. Briggs and M.P. Seah, editors, *Practical Surface Analysis by XPS and AES*, (Wiley, Chichester, 1983).

[20] W. L. Baun, Surf. Interface Anal. **3**, 243 (1981).

[21] S. A. Benninghoven, F. G. Rudenauer and H. W. Werner, *Secondary Ion Mass Spectrometry*, (Wiley, New York, 1987).

[22] H.-K. Jeong, T. Komesu, C.-S. Yang, P. A. Dowben, B. D. Schultz and C. J. Palmstrøm, Mater. Lett. **58**, 2993 (2004).

[23] M.P. Seah, M.E. Jones and M.T. Anthony, Surf. Interface Anal. **6**, 242 (1984).

[24] M.P. Seah, Surf. Interface Anal. **20**, 243 (1993).

[25] S. Tanuma, S. Ichimura, and K. Yoshihara, Appl. Surf. Sci. **101**, 47 (1996).

[26]   NIST Electron Inelastic Mean Free Path program, ver. 1.1 (NST std. ref. Database 71) http://www.nist.gov.srd

[27]   P.A. Dowben, M. Grunze and D. Wright, Surf. Sci. **134**, L524-L582 (1983).

[28]   U. Vahalia, P.A. Dowben and A. Miller, J. Electron Spectrosc. Rel. Phenom. **37**, 303 (1986).

[29]   S. Tamura, C.J. Powell, and D.R. Penn, Surf. Interface Anal. **17**, 911 (1991).

[30]   S. Tamura, C.J. Powell, and D.R. Penn, Surf. Interface Anal. **21**, 165 (1994).

[31]   C.J. Powell and A. Jablonski, J. Vac. Sci. Technol. **A17**, 1122 (1999).

[32]   G. Schönhense and H.C. Siegmann, Ann. Phys. (Leipzig) **2**, 465 (1993).

[33]   H.-J. Drouhin, Phys. Rev. B **56**, 14886 (1997).

[34]   M. Lundwall, M. Tchaplyguine, G. Öhrwall, A. Lindblad, S. Peredkov, T. Rander, S. Svensson, and O. Bjørneholm, Surface Sci. (2005) in press

[35]   D. Spanjaard, C. Guillot, M.-C. Desjonqueres, G. Treglia, and J. Lecante, Surf. Sci. Repts. **5**, 1 (1985).

[36]   R. Monnier, Phil. Mag. B **75**, 67-144 (1997).

[37]   S.M. Foiles, Phys. Rev. B **32**, 7685 (1985).

[38]   M.W. Finnis and J.E. Sinclair, Phil. Mag. **A50**, 45 (1984).

[39]   G. Bozzolo, J. Ferrante, R.D. Noebe, B. Good, F.S. Honecy and P. Abel, Comput. Mater. Sci. **15**, 169 (1999).

[40]   H. Deng, W. Hu, X. Shu and B. Zhang, Appl. Surf. Sci. **221**, 408 (2004).

[41]   J.L. Moran-Lopez, Surf. Sci. Lett. **188**, L472 (1987).

[42]   A.N. Caruso and C.N. Borca, in *Recent Developments in Magnetism and Magnetic Materials* **1**, pp. 33-48 (2003).

[43]   S. Y. Liu and H. H. Kung, Surf. Sci. **110**, 504 (1981).

[44]   A. Knop, D.N. McIlroy, P.A. Dowben, and E. Rühl, in *Two-Center Effects in Ion-Atom Collisions*, T. J. Gay and A.F. Starace editors, (*AIP Conference Proceedings* **362**, 1996) pp. 274-280.

[45]   B.Yang, M.Asta, O.N. Mryasov, T.J. Klemmer and R.W. Chantrell, Scripta Materialia **53**, 417-422 (2005).

[46]   R.V. Chepulskii, J. Velev and W.H. Bulter, J. Appl. Phys. **97**, 10J311 (2005).

[47]   Natalie Palina, H, Modrow, R. Müller, J. Hormes, P.A. Dowben and Ya.B. Losovyj, Materials Letter, (2005) in press.

[48]   S. Kurisu, T. Ido and H. Yokoyama, IEEE Trans. Magn. **MAG-23**, 3137 (1987).

[49]   A. Miller and P.A. Dowben, J. Phys. Cond. Matt. **5**, 5459 (1993).

Mater. Res. Soc. Symp. Proc. Vol. 887 © 2006 Materials Research Society 0887-Q08-01

Size Control and Spectroscopic Characterization of Monolayer Protected Gold Nanoparticles Obtained by Laser Ablation in Liquids

Giuseppe Compagnini, Alfio Alessandro Scalisi and Orazio Puglisi[*]

Dipartimento di Scienze Chimiche, Università di Catania, Viale A. Doria 6 Catania 95125

Italy

ABSTRACT

In this paper we present a study on the formation of gold colloids by laser ablation of a gold metal target in alkanes and thiol-alkane solutions. The results show a decrease of the gold particles' size up to 2 nm when thiol molecules are present in the liquid environment. In summary, we observed that laser ablation of gold targets in thiol-alkane solutions leads to the formation of stable gold clusters with size smaller than those obtained in the corresponding pure alkane. This result is a consequence of the competition between the aggregation of gold species in the plume (which allows a gold embryo to be formed and to grow) and the tendency of the dispersed thiol molecules to bond at each embryo surface stopping their growth.

INTRODUCTION

Formation of metal nanoparticles by laser ablation in liquid phase has been reported in several recent papers[1-9]. Actually this methodology offers some advantages with respect to more traditional synthetic routes such as the chemical methods and laser ablation in vacuum. The most important advantage against the chemical synthetic routes is the absence of toxic products that could be dangerous for the man health and for the environment. [10]. With respect to laser ablation the most beneficial advantage is the absence of any vacuum requirement which in turn means simplicity and high productivity with respect to any possible engineering application.

Recently, we have shown that this method allows to obtain colloids in several different liquid solvents like water, ethanol, and various alkanes,[11] enhancing the practical use of the obtained sols for industrial purpose. Particular attention has been devoted in our laboratory to the production of silver and gold sols owing to the many applications of this material, but the method is in principle applicable to any material that can be ablated. Among the problems related to the formation of metal nanoparticles by laser ablation in liquids we mention the post

---

[*] Corresponding Author: Orazio Puglisi, Tel. and Fax +39095221635; E-mail: opuglisi@unict.it

ablation re-aggregation processes. These phenomena in turn force the particles to rapidly coalesce with the result of a broad cluster size distribution. In this respect, , Mafuné et al.(1–3,12-14), in their pioneering works demonstrated that the cluster size can be drastically reduced by the use of aqueous solution of surfactants, which cover the particles just after their ablation and thus prevent further agglomeration. (12-14). In particular, when sodium dodecyl sulfate (SDS) is used, the size distribution of gold, silver, and platinum particles tends to become narrower with increasing surfactant concentration. However, this method is strongly limited to the water environment, and in particular to a two-phase system (micelle-water). We recently found evidence that one can obtain gold nanoparticles in homologous series of alkanes from C5 to C10 (pentane to decane), thus controlling size and shape of the particle as a function of ablation environment.(15). In this paper we want to report some data on the ablation in alkane-alkanethiol solutions. These experiments not only show the possibility to reduce the size of the ablated gold particles in non aqueous media but will demonstrate that a central issue in this phenomenology is represented by the surface chemistry of the nanostructures. In particular fine details of the surface chemistry can provide the key factor against agglomeration and coagulation of the sol.

EXPERIMENTAL DETAILS

Ablation has been conducted using a 532-nm radiation coming from a Nd-yttrium aluminum garnet (YAG) laser with a 5-ns pulse duration and a 10-Hz repetition rate. An optically smooth gold plate was cleaned in an ultrasonic bath using several solvents to remove organic contaminations and it was placed at the bottom of a glass vessel filled with the chosen liquid. The thickness of the liquid above the target was about 15 mm, and the ablation time was maintained at 5 min. The laser beam was focused on the target using a 10-cm focal lens placed above the vessel. This gives the possibility to change the laser fluence (between 0.5 and 30 J/cm2) by changing the lens target distance. Several alkanes (from pentane to decane) pure and in dodecanethiol [$CH_3(CH_2)_{11}SH$ (DDT)] solutions have been considered as ablation environments. The infrared spectra have been taken using a Bruker Equinox 55 Fourier Transform machine, working in the range of 400-4000 cm^{-1} with a spectral resolution of 2 cm^{-1}. Electron micrographs of Au nanoparticles were obtained by using a transmission electron microscope (JEOL JEM-2010F).

DISCUSSION

Thiol capped gold nanoparticles have been obtained by ablating a gold metal target into different decane-dodecanethiol (DDT) solutions. Details on the preparation and the particles' structure as function of the experimental parameters can be found elsewhere (11,15). Once the ablation starts, it is possible to observe the formation of the colloid because after a few seconds, the solution gradually turns brownish-red with a colour intensity, which increases with increasing ablation time. Colloids obtained in DDT-alkane solutions were found to be very stable for several weeks,

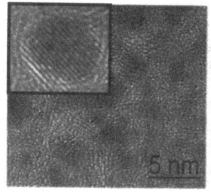 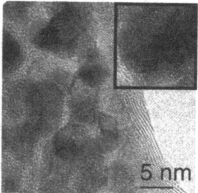

FIG. 1. *TEM images for Au nanoparticles obtained (left) in a DDT solution 0.01 M and (right) in n-decane. In insets Au nanoparticles are shown to be single crystalline and polycrystalline, respectively.*

maintaining, unaltered, their intense purple colour, their plasmon resonance features and the size distribution, as measured by Transmission Electron Microscopy, while those obtained in pure alkanes show agglomeration processes after a few days and a visible change of their colour.

The average particle size has been found to decrease by increasing the thiol concentration, starting from 6 nm (pure decane) down to 1.6 nm in the case of a thiol concentration around 0.05 M. The width of the distribution is also affected by the thiol concentration spanning from 4 nm in the first case to 2 nm in the case of the highest concentration. Some drops of the colloids have been also deposited on suitable substrates for the spectroscopic investigations. Interesting insights come from transmission electron microscopy (TEM) of the deposited gold particles. Figure 1 reports TEM images for particles obtained in pure decane [right] and in a decane-DDT solution [left]. In this case, ablation has been performed at 1 J/cm2 in a 0.01 M dodecane thiol solution. It is clear that laser ablation in these systems produced gold crystalline clusters, most of them showing a spherical shape. However, particles obtained in the DDT decane solution were generally single crystalline and well separated [see inset in Fig. 1], while in the case of ablation in pure decane they show grain boundaries suggesting a particle agglomeration.

The overall effect of DDT molecules in the formation and evolution of gold sols can be accounted with arguments similar to those given by Mafuné et al.(1–3,12–14). After the laser pulse onto the metallic target, there is a rapid formation of an embryonic metallic particle followed by particle growth in which, at the particle's surface, there is competition between the growth and the termination due to surfactant coating onto the particle. According to the Mafune's model, following the laser shot, a dense cloud of gold atoms (plume) is built over the laser spot. This cloud refurnishes the growing surface until the atoms in close vicinity are depleted, leaving the embryonic nanoparticles surrounded by a region void of the atoms.

Further growth of the particle is now limited by the diffusion of atoms from outside the void and is slow. Moreover, diffusion of DDT molecules towards the growing particle and coating of the surface gives rise to further slowing of the growth rate until termination occurs when all the surface is covered with DDT molecules. Indeed, the presence of a chemical agent (DDT molecules) which is able to have strong dative bonds with the metal can slow down the particle growth because of the competitive S-Au bond formation. In addition, the coating of gold particles by an organic alkyl shell prevents further agglomeration and stabilizes the colloid. The role of the dodecanethiol is then to freeze the embryonic particles produced during the ablation process. Preliminary infrared absorption and x-ray photoelectron spectroscopy data have shown that the alkyl chain packaging around each Au particle is liquid-like ordered as the result of the dynamic process of gold aggregation during ablation process and the reduction of the effective coordination number in small nanoparticles.

Samples have been also prepared in order to investigate their infrared absorption due to the organic shell. Two regions in the absorption spectrum are important in our study: in the first, S-H stretching modes are observed at around 2570 cm^{-1}. These have been found absent in all the investigated samples, confirming the chemisorption of all thiol molecules with the formation of covalent Au-S bonds and disappearance of the SH bond. (16). In the IR spectrum the C-H stretchings of alkyl chains are observed between 2800 and 3100 cm^{-1} (see Fig.2).

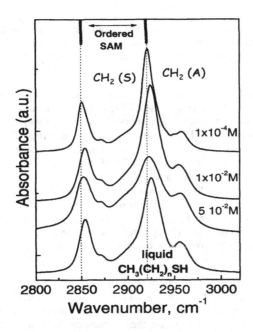

Fig. 2 *Infrared absorption in the CH stretching range for several samples. These data are compared with those coming from a perfectly ordered 2D-SAM on a gold (111) surface and a disordered system such as a liquid thiol.*

This spectral region deserves attention because it can show size effect on the vibrational response. Size effects can be forecasted in the arrangement of the alkyl capping chains of the nanoparticles. Two effects are relevant in comparing self assembly on flat metal surfaces (bidimensional Self Assembled Monolayers:2D-SAM) with that on finite size metal particles (3D-SAM). These are: (i) the increase in concentration of the surface active bond sites as the particle size is reduced and (ii)   the finite radius of curvature. 2D-SAM systems like those on flat Au (111) substrates possess all the chemisorbed thiol molecules well ordered with mostly all methylene groups in trans zigzag configuration (17). On the other hand a 3D-SAM system, like that on a small cluster of around 2 nm, shows a less ordered thiol chain packaging (especially for long chain lengths) with many gauche defects at the chain endings (18,19). A carefully study of the vibrational features of these chains can be used to evidence this effect. Fig. 2 reports the result for such a study on the CH stretching region. In this range five signals are generally considered. Two of them are related to CH_2 stretchings (symmetric at 2850 cm^{-1} termed d+ and asymmetric at 2920 cm^{-1} termed d-) and other two are assigned to CH_3 ones (symmetric at 2870 cm^{-1} and asymmetric at 2960 cm^{-1}). Finally a structure around 2900 cm^{-1} is attributed to a CH_3 Fermi resonance[16]. In a conformation study, special emphasis is given to d+ and d- bands because of their correlation to the average trans and gauche methylene population within the chains. This feature is then generally used to evaluate the order of a SAM film with a red-shift of about 5-6 cm^{-1} by increasing order. In our case such an effect is evident by changing the average particle size and comparing the results with 2D-SAM and liquid thiol spectra. In fact, while samples with average particles of 4 nm (or above) show the CH_2 stretching positions at 2850 and 2920 cm^{-1} (d+ and d- in an ordered SAM), in the case of the sample having a 1.6 nm average particle size these positions shift to 2854 and 2924 cm^{-1}, that is very close to the ones observed for a liquid thiol (2855 and 2925 cm^{-1}). This effect reveals that with this size chain disorder is almost complete, due mainly to the surface curvature. At the same time an increase of the peak widths is observed in agreement with literature [20]. In any case, a monotone decrease of the I(d-)/I(d+) integrated intensity ratio with decreasing the particle size confirms the angle distortion with respect to the ideal assembly on flat surfaces.

References

1. F. Mafuné, J. Kohno, Y. Takeda, T. Kondow, and H. Sawabe: Formation and size control of silver nanoparticles by laser ablation in aqueous solution. *J. Phys. Chem. B* **104,** 9111 (2000).

2. F. Mafuné, J. Kohno, Y. Takeda, T. Kondow, and H. Sawabe: Structure and stability of silver nanoparticles in aqueous solution produced by laser ablation. *J. Phys. Chem. B* **104,** 8333 (2000).

3. F. Mafuné, J. Kohno, Y. Takeda, T. Kondow, and H. Sawabe: Formation of gold nanoparticles by laser ablation in aqueous solution of surfactant. *J. Phys. Chem. B* **105,** 5114 (2001).

4. T.Sakai, Y.Takeda, F.Mafuné, M. Abe, and T. Kondow: Monitoring growth of surfactant-free nanodroplets dispersed in water by single-droplet detection. *J. Phys. Chem. B* **107,** 2921 (2003).

5. F.Mafuné, J.Kohno, Y.Takeda, and T.Kondow: Nanoscale soldering of metal nano particles for construction of higher-order structures. *J. Am. Chem. Soc.* **125**, 1686 (2003).

6. A.V. Simakin, V.V. Voronov, G.A. Shafeev, R. Brayner, and F. Bozon-Verduraz: Nanodisks of Au and Ag produced by laser ablation in liquid environment. *Chem. Phys. Lett.* **348**, 182 (2001).

7. S.I. Dolgaev, A.V. Simakin, V.V. Voronof, G.A. Shafeev, and F. Bozon-Verduraz: Nanoparticles produced by laser ablation of solids in liquid environment. *Appl. Surf. Sci.* **186**, 546 (2002).

8. T. Tsuji, K. Iryo, Y. Nishimura, and M. Tsuji: Preparation of metal colloids by a laser ablation technique in solution: Influence of laser wavelength on the ablation efficiency (II). *J Photochem. Photobiol. A* **145**, 201 (2001).

9. A.V. Kabashin and M. Meunier: Synthesis of colloidal nanoparticles during femtosecond laser ablation of gold in water. *J. Appl. Phys.* **94**, 7941 (2003).

10. A.V. Kabashin, M. Meunier, C. Kingston, J. H.T. Luong, J. Phys. Chem. B 107 (2003) 4527.

11. G. Compagnini, A.A. Scalisi, O. Puglisi, Phys. Chem. Chem. Phys. 4 (2002) 2787

12. F. Mafuné and T. Kondow: Formation of small gold clusters in solution by laser excitation of interband transition. *Chem. Phys.Lett.* **372**, 199 (2003).

13. F. Mafuné, J. Kohno, Y. Takeda, and T. Kondow: Formation of stable platinum nanoparticles by laser ablation in water. *J. Phys.Chem. B* **107**, 4218 (2003).

14. F. Mafuné, J. Kohno, Y. Takeda, and T. Kondow: Dissociation and aggregation of gold nanoparticles under laser irradiation.*J. Phys. Chem. B* **105**, 9050 (2001).

15. G. Compagnini, A.A. Scalisi, and O. Puglisi: Production of gold nanoparticles by laser ablation in liquid alkanes. *J. Appl. Phys.* **94**, 7874 (2003).

16. A. Manna, T. Imae, T. Yogo, K. Aoi, M. Okazaki, J. Colloid Interf. Sci. 256 (2002) 297.

17. M.D.Porter, T.B. Bright, D.L. Allara, C.E.D. Chidsey, J. Am. Chem. Soc. 109 (1987) 3559.

18. T. Pham, J.B. Jackson, N.J. Halas, T.R. Lee, Langmuir 18 (2002) 4915.

19. P. Jiang, S-S. Xie, J-N. Yao, S-J Pang, and H-J. Gao, J. Physics D: Appl. Phys. 34 (2001) 2255. [19] C.K. Yee, A. Ulman, J.D. Ruiz, A. Parikh, H. White, M. Rafailovich, Langmuir 19 (2003) 9450.

20. C.K. Yee, A: Ulman, J.D. Ruiz, A. Parikh, H. White, M. Rafailovich, Langmuir 19 (2003) 9450

Mater. Res. Soc. Symp. Proc. Vol. 887 © 2006 Materials Research Society   0887-Q05-03

Electron work function at grain boundary and the corrosion behavior of nanocrystalline metallic materials

D.Y. Li
Dept. of Chemical and Materials Engineering
University of Alberta, Edmonton, Alberta, Canada T6G 2G6
dongyang.li@ualberta.ca

ABSTRACT

Due to their high grain boundary density, nanocrystalline materials possess unusual mechanical, physical and chemical properties. Extensive research on nanocrystalline materials has been conducted in recent years. Many studies have shown that corrosion, one of important properties of nanocrystalline materials, is crucial to their applications. In this article, the activity of electrons at grain boundaries of metallic surfaces is analyzed based the electron work function (EWF), the minimum energy required to attract electrons from inside a metal. It is demonstrated that at grain boundaries, the electron work function decreases, indicating that at a grain boundary, electrons are more active. As a result, the surface becomes more electrochemically reactive. Such increase in electrochemical reactivity has negative effect on the corrosion resistance of nanocrystalline materials. However, for a passive nanocrystalline metal or alloy, the nanocrystalline structure is beneficial to its corrosion resistance through rapid formation of a protective passive film. The mechanisms responsible for the variation in EWF at grain boundary and effects of nanocrystallization on corrosion are discussed in this article.

INTRODUCTION

In recent years, the nanocrystalline structure has attracted extensive interest, since it possesses properties that are different from those of conventional polycrystalline structures having their grain size at the micron level. The high-density grain boundaries in nanocrystalline materials considerably vary their intrinsic properties. For instance, nanocrystalline Ni shows lower resistance to corrosion than regularly-grained Ni [1], while nanocrystalline stainless steel performs better than regularly-grained stainless steel [2]. It is known that the corrosion resistance of a metal is largely dependent on its surface electron activity. The more active the surface electrons, the higher the corrosion rate [3]. It is therefore of interest to investigate effects of grain boundary on the electron behavior in order to gain an insight into the corrosion behavior of nanocrystalline materials. The electron behavior may be characterized by the electron work function, which is the minimum energy required to remove an electron from an electrically neutral solid in vacuum [4]. Recent studies have shown that electrons at grain boundaries are more active than those inside grains and the electron work function decreases with the grain boundary density [5]. In this article, the author summarizes results of recent studies conducted in his Surface/Tribology Lab on the work function, corrosion, and nanocrystalline materials, based on which the correlation between the electron behavior and corrosion of nanocrystalline materials is discussed. Efforts are made to elucidate the relationships among EWF, grain size, grain boundary, and electrochemical properties.

EXPERIMENTAL STUDIES AND DISCUSSION

- Effects of grain boundary on EWF and corrosion behavior

Effects of grain boundary on EWF were previously investigated using copper and aluminum as sample materials [5]. Cast Cu and Al samples ($10 \times 10 \times 10mm$) were annealed at $800\,°C$ and $550\,°C$, respectively, for 1 hour. The treated copper and aluminum samples had their average grain sizes of 2.5 mm and 2.0 mm, respectively. In order to determine how the grain size influenced the overall EWF, copper samples were cold-rolled and annealed at $750\,°C$ for different periods of time, leading to various grain sizes ranged from 0.04 mm to 0.6 mm. All the copper and aluminum samples were polished using abrasive sand papers first and then 0.05 μm colloidal silica, followed by etching with a $FeCl_3$+HCl solution and a HF solution, respectively, for 5 seconds to eliminate the deformed layer caused by polishing. The samples were finally cleaned using an ultrasonic cleaner with reagent grade acetone (for 10 minutes) and reagent alcohol (for 5 minutes), and then dried in an inert container.

Electron work functions (EWFs) of the samples were measured using a scanning Kelvin probe with a gold tip, provided by KP Technology (Caithness, UK) [14]. A three-axis micro-stepper positioner permitted high-resolution sample positioning (0.4 μm/step). The spacing between the probe and the specimen surface could be controlled within 40 nm. More details about the EWF measurement have been given in reference [5].

Fig.1 illustrates variations in EWF as the Kelvin probe scanned over a distance that covered a number of grain boundaries. As shown, the EWF dropped when the probe moved across a grain boundary (GB) or a triple point (TP) at which three GBS met. Such a drop of EWF indicates that electrons at grain boundary are more active.

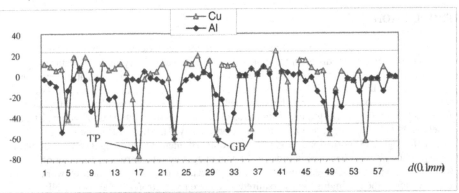

Figure1. The EWF dropped as the Kelvin probe moved across grain boundaries [5].

The increase in electron activity at grain boundary implies that the surface reactivity of surfaces having smaller grain sizes would be higher. This has been demonstrated by measuring EWFs of copper samples having different grain sizes. As shown in Fig.2, a decrease in the grain size led to a decrease in EWF, corresponding to an increase in the surface reactivity.

A surface with higher reactivity should be less resistant to corrosion. This has been shown by recent experimental studies on nanocrystalline and microcrystalline copper electrodeposits (see Fig.3) produced using pulse-current and direct-current processes, respectively [6]. Polarization tests demonstrated that the nanocrystalline Cu deposit (grain size: 56 nm) had a higher corrosion current than the microcrystalline deposit (grain size: $2\,\mu m$) in a 1 $M H_2SO_4$ solution, in which copper was not passivated. The corrosion currents of the deposits are consistent with their electron work functions as shown in Table I (Samples were immersed in the H_2SO_4 solution for two hours, then rinsed with acetone and dried before the EWF measurement).

Table I. Average grain size, corrosion current and EWF of nanocrystalline and Microcrystalline Cu electrodeposits

Electrodeposit	Grain size (nm)	$i_{corr}\,(\mu A/cm^2)$; 1 $M H_2SO_4$	EWF (eV)
Nano-Cu	56	24.0	4.15
Micro-Cu	2000	16.0	4.19

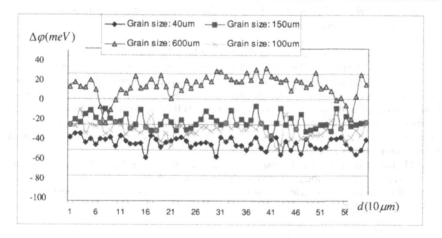

Figure 2. EWF varied as the Kelvin probe moved on surfaces of copper samples having different grain sizes [5]. d is the displacement of the probe and $\Delta\varphi$ is the change in EWF.

The effect of grain boundary on the activity of electrons at grain boundary is understandable. The grain boundary may be viewed as a dislocation wall. For instance, a low-angle tilt grain boundary is an array of parallel edge dislocations [7]. For a high-angle grain boundary, it may still be regarded as a dislocation wall that has higher-density mixed dislocations. It has been demonstrated that electrons around a dislocation are more active and easier to be removed from the region [8]. Therefore, the electron work function at grain boundary should be lower than that inside a grain. As a result, a metallic surface should have a higher dissolution rate or a larger corrosion current than a surface having a lower grain boundary density. This has been demonstrated by different corrosion currents of the above-mentioned nanocrystalline and microcrystalline Cu deposits in the H_2SO_4 solution (Table I).

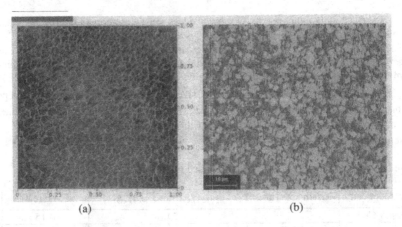

(a)   (b)

Figure 3. (a) A nanocrystalline Cu electrodeposit (average grain size: 56 nm) and (b) a microcrystalline Cu electrodeposit (average grain size: $2\,\mu m$), made using pulse-current and direct-current processes, respectively [6].

It should be noted that the decrease in EWF could be influenced by the change in surface geometry at a grain boundary. After etching, a grain boundary may appear as a micro groove, whose size depends on the degree of etching, when viewed under a microscope. A groove could increase the local electron activity, since electrons at angular edges of a groove could be more active due to a recent study of the roughness effect on EWF [15]. However, the dislocations at grain boundary could play a predominant role in corrosion, since the angular edges may become blunt and thus decrease the local corrosion rate as corrosion proceeds if the active electrons mainly come from regions around the angular edges.

It should also be mentioned that passive films might form on the Cu and Al samples after etching, which may influence their EWF as well. However, such formed passive films could be too thin to block electrons to escape from the metals, since the EWF still dropped at grain boundaries as Fig.1 illustrates.

- EWF and the corrosion behaviour of nanocrystalline passive materials

However, the above-mentioned conclusions do not apply to nanocrystalline passive materials or the cases when passivation occurs during corrosion. Fig.4 illustrates polarization curves of the nanocrystalline and microcrystalline Cu electrodeposits in a 0.1M *NaOH* solution, in which the metal can be passivated. Based on the electrochemical measurement, the corrosion currents were determined using the Stern-Geary relation [13]. Table II provides the information on corrosion currents of the deposits and their EWFs measured after being immersed in the *NaOH* solution for two hours (rinsed with acetone and dried before the EWF measurement).

Clearly, the nanocrytsalline Cu deposit is more resistant to corrosion in this solution involving passivation. Such a positive effect of nanocrystallization on corrosion behavior should be attributed to the formation of a protective passive film that can be enhanced by nanocrystallization of the substrate. This mechanism has also been demonstrated by a study on

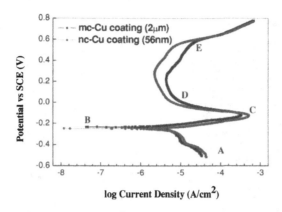

Figure 4. Dynamic polarization curves of nanocrystalline and microcrystalline Cu deposits in a 0.1M *NaOH* solution [6].

Table II. Average grain size, corrosion current and EWF of nanocrystalline and microcrystalline Cu electrodeposits [6]

Electrodeposit	Grain size (nm)	$i_{corr}(\mu A/cm^2)$; 0.1M *NaOH*	EWF (eV)
Nano-Cu	56	3.6	4.62
Micro-Cu	2000	8.8	4.33

EWF and corrosion of nanocrystalline 304 stainless steel (304SS) [9]. In the study, stainless steel samples with nanocrystalline surfaces were made by sandblasting and annealing [9]. The sandblasting resulted in a heavily deformed surface layer with dislocation sub-cells and diffuse sub-cell boundaries. The annealing treatment (recovery treatment at $350°C$) sharpened the sub-grain boundaries and resulted in nanocrystallites having an average grian size in the range of 20 ~ 30 nanometers (Fig.5 (a)). Fig.5(b) illustrate polarization curves of three different 304 steel samples: surface nanocrystallized, sandblasted without the recovery treatment, and regaularly-grained (average grain size: ~ 50 *μm*). One may see that the nanocrystalline surface had superior polarization behavior in comparison with the regularly-grained one. The sandblased one showed the poorest performance due to its high-density dislocations.

The superior polarization behavior of nanocrystalline surface of 304 stainless steel could be mainly attributed to the formation of an improved passive film with enhanced protective capability against corrosion. An improved passive film renders the surface more electro-chemically stable with a higher EWF. Fig. 6 illustrates EWFs of the three surfaces. Clearly, the nanocrystalline surface had higher EWF than the regularly-grained one, indicating that the passive film on the nancrystalline surface made it more electrochemically stable against electrochemical attack. Such enhanced resistance to corrosion may largely benefit from improved mechanical behavior of the formed passive film. In order to evaluate mechanical properties of the passive film in comparison with that on the regularly-grained sample, nanoindentation tests were performed on surfaces of the three samples. Fig.7 illustrates force –

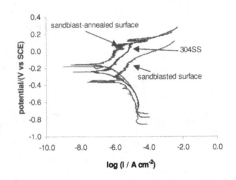

(a)

Figure 5. (a) TEM micrograph of a nanocrystalline surface generated by sandblasting and recovery treatment; (b) the surface nanocrystalline stainless steel showed the best polarization behavior, followed by the regularily-grained stainless steel (average grain size: ~ 50 μm), and the sandblasted one showed the poorest performance [9]. Solution: 3.5% $NaCl$.

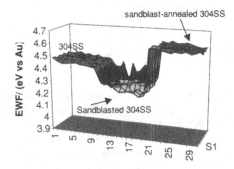

Figure 6. EWFs of nanocrystalline, regularly-grained, and sandblasted stainless steel surfaces [9].

depth curves of the three samples. One may see that the the passive film on the nanocrystalline surface was more elastic and harder (corresponding to a smaller indentation depth) in comparison with that on the regaularly-grained sample. The passive film on the sandblasted sample was the poorest in terms of the mechanical behavior. Scratch resistances of the three passive films were also determined by performing micro-scratch tests with monitoring changes in the contact electric resistance (CER), which more or less reflected the adherence of a passive film to its substrate. During such a test, a target surface was scratched by a diamond tip under a continuously increased normal force and corresponding changes in CER was recorded. When the passive film was damaged, a corresponding drop in CER occurred. The critical load under which the passive film failed was used as a parameter to evaluate the scratch resistance of the passive film. Fig.8 illustrates representative changes in load and CER with respect to time of scratching. As shown in Table III, the passive film on the nanocrystalline surface had the highest

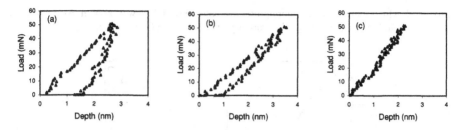

Figure 7. Load ~ depth curves of passive films under a maximum load of 50μN. (a) 304 SS, (b) Sandblasted 304SS and (c) Sandblast-annealed 304SS [9].

resistance to scratch, followed by the regularly-grained one, and that on the sandblasted one was the poorest.

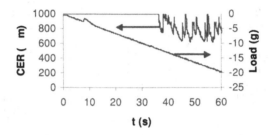

Figure 8. Typical changes in the CER with an increase in load against time for the passive film on a nanocrystalline 304 SS surface during micro-scratch [9].

Table III. Critical loads* under which passive films failed during micro-scratching [9]

Specimens	304SS	Sandblasted 304SS	Sandblast-annealed 304SS
Critical load (g)	7.5 ±1.7	5.5 ±1.2	12.5±3.0

* Each value was obtained by average 3 measurements

The improvement in properties of the passive film could be attributed to promoted diffusion of Cr atoms in the nanocrystalline 304 stainless steel. When the grain size was reduced to the nanometer scale, the atomic diffusion could be considerably accelerated not only along grain boundaries but also inside grains, since a nanocrystalline crystal lattice is markedly distorted with residual strain [10], which favours atomic diffusion and could thus dramatically influence the formation of passive film and its properties. Fig.9 illustrates signals from the nanocrystalline surface and that of the regularly-grained one during XPS analysis. It was demonstrated that the count of chromium from the nanocrystalline surface was higher than that from the regularly-grained sample. This result may imply that the diffusion of chromium towards surface was promoted, which may result in a more protective passive film. It is well known that increasing the Cr content in a Cr-containing alloy can improve its passive film [11, 12].

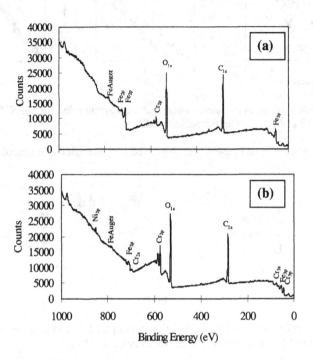

Figure 9. XPS spectra of natural passive films respectively on (a) regularly-grained 304 stainless steel and (b) surface-nanocrystallized 304 stainless steel.

CONCLUSIONS

1. Electron work function decreases at grain boundary, indicating that electrons at grain bounadry are more active, which make the grain bounadry vulnerable to electrochemical attack.

2. Consequently, a nanocrystalline material may have lower resistance to corrosion than its regularly-grained counterpart.

3. However, if a material is passive, it would be more resistant to corrosion with an enhanced passive film than its regularly-grained counterpart. The promoted atomic diffusion in the nanocrystalline material could be mainly responsible for such improvement.

ACKNOWLEDGMENTS

The author would like to thank the following research fellows in the Surface/Tribology Lab for their great contribution to this study: Drs. X.Y. Wang, X.S. Guan, Z.F. Dong and W. Li (former postdoctoral fellows) and Tao Song (M.Sc. student). The research is sponsored by Natural Science and Engineering Research Council of Canada (NSERC) and Alberta Science and Research Authority (ASRA).

REFERENCES

1. Rofagha et al, The corrosion behavior of nanocrystalline nickel, *Scripta Metall. Mater.*, **25**, 2867 (1991).
2. R.B. Inturi and Z. Szklarska-Smialowski, Localized corrosion of nanocrystalline 304 type stainless steel films, *Corrosion*, 398 (1992).
3. W. Li and D.Y. Li, Variations of work function and corrosion behaviors of deformed copper surfaces, *Appl. Surf. Sci.*, **240**, 388 (2005).
4. Michael P. Marder, Condensed Matter Physics, John Wiley & Sons, Inc., New York, 2000, PP. 519-520.
5. W. Li and D.Y. Li, Effect of grain size on the electron work function of Cu and Al, *Surf. Rev. and Lett.*, **11**, 173 (2004).
6. S. Tao and D.Y. Li, Tribological, mechanical and electrochemical properties of nanocrystalline copper deposits produced by pulse electrodeposition, *Nanotechnology*, **17**, 65 (2006).
7. D.A. Porter and K.E. Easterling, Phase transformation in metals and alloys, 2nd edition, Chapman & Hall, London, 1992, Chapter 3.
8. W. Li and D.Y. Li, Effects of Dislocation on the Electron Work Function of a Metal Surface, *Mater. Sci. & Tech.*, **18**, 1057 (2002).
9. X.Y. Wang and D.Y. Li, Mechanical, electrochemical and tribological properties of nanocrystalline surface of 304 stainless steel, *Wear*, **255**, 836 (2003).
10. X.S. Guan, Z.F. Dong and D.Y. Li, Surface nanocrystallization by sandblasting and annealing for improved mechanical and tribological properties, *Nanotechnology*, **16**, 2963 (2005).
11. D. Hamm, C. Olsson, & D. Landolt, Effect of chromium content and sweep rate on passive film growth on iron-chromium alloys studied by EQCM and XPS, *Corro. Sci.*, **44**, 1009 (2002).
12. S. Fujimoto, T. Shibata, & T. Yamada, Suppression of pitting corrosion with passive film modification on type 304 stainless steel by ultra-violet light irradiation, *J. of the Japan Institute of Metals*, **62**, 527 (1998).
13. Stern M, Geary A L, *J. Electrochem. Soc.*, **104**, 56 (1957).
14. I. D. Baikie, P. J. S. Smish, D. M. Porterrfield, and P. J. Estrup, Multitip scanning bio-Kelvin probe *Rev. Sci. Instrum.*, **70**, 1842 (1999).
15. W. Li and D.Y. Li, On the correlation between surface roughness and work function in copper, J. of Chem. Phys., **122**, 064708 (2005).

Mater. Res. Soc. Symp. Proc. Vol. 887 © 2006 Materials Research Society   0887-Q06-10

Sintering of HAp precipitated from solutions containing ammonium nitrate and PVA

Tatiana V. Safronova[1], Valery I.Putlayev[1], Alexey V.Belyakov[2], Mikhail A. Shekhirev[3]
[1]Chemical Department and [3]Department of Materials Science of Moscow State University, Leninskie Gory, 1, Moscow 119992, Russia
[2]Department of Chemical Technology of Silicates of Mendeleev Chemical Technological University of Russia,
Miusskaya pl., 9, Moscow 125047, Russia

ABSTRACT

Bioceramics based on hydroxyapatite (HAp) is known to be a prospective material for biomedical applications. However, sintering of HAp is still understudied in sense of reasonable selection of controlling parameters. In particular, the role of impurities and co-products of powder fabrication is still questionable. The data concerning the role of ammonium nitrate coming to precipitated HAp from the mother liquor, its effect on powder compaction and subsequent sintering, are not available.

Nanosized powders of HAp were fabricated via conventional wet-precipitation technique by dropwise adding of $Ca(NO_3)_2$ solution (0.25 -1.67 M) to the stock solution of $(NH_4)_2HPO_4$ (0.15-1.00 M) with pre-adjusted pH at 60 C in presence of polyvinyl alcohol (PVA). PVA was added to the stock solution in order (i) to block crystal growth during synthesis , (ii) to improve stability of Hap suspension to sedimentation, (iii) to regulate an aggregation of HA nanoparticles during synthesis and in the stage of drying.

NH_4NO_3 – a by-product of the precipitation reaction, presented in as-precipitated powder in amount of 30%, was evaluated as an additive affecting a compaction of the powder and the initial stage of a sintering. The powder samples were tested by XRD, FTIR, light-scattering , TEM and SEM/EDX to get particle sizes, morphology and chemical composition, dilatometry. Ceramics were sintered at 700-1250 C and evaluated with SEM/EDX and density measurements.

Addition of PVA to the stock solution in the course of HAp precipitation is a promising technique to control an aggregation of HAp nanoparticles in the stages of drying and sintering. PVA acting as a surfactant in the solution and as a binder in dry powder can keep highly reactive small HAp particles within large agglomerates providing better molding of the powder and controllable densification of ceramics. The effect of PVA on microstructure of the HAp powder and their sintering behaviour is discussed in terms of self-organisation concept and synergetics.

INTRODUCTION

Ceramics based on calcium phosphates is known to be a prospective material for biomedical applications [1]. Much attention is paid to hydroxyapatite $Ca_{10}(PO_4)_6(OH)_2$ (HAp) due to its affinity to a bone mineral. Different precursors have been used to synthesize HAp powders. The most popular reaction consists in precipitation from $(NH_4)_2HPO_4$ and $Ca(NO_3)_2$ solutions. Effect of the principal parameters of the powder fabrication on sintering was extensively discussed (see, e.g., [2]). However, the role of impurities and co-products of powder fabrication is still questionable. Thus, there is no data concerning the role of ammonium nitrate coming to precipitated HAp from the mother liquor, its effect on powder compaction and subsequent sintering, though NH_4NO_3-monitoring of wet precipitates and bulk samples prior to sintering is considered as a routine [3]. Meanwhile, NH$_4$NO$_3$, a low-melting impurity, could cause a beneficial effect on a compaction of the powder and on the initial stage of sintering. Another key is a role of organic (mostly, polymeric) substances acting in the course of HAp preparation. This subject was addressed significant amount of works pointing out a positive influence of these additives, especially surfactants [4-6] which can decrease surface energy due to adsorbtion or decrease aggregation due to decreasing of drying temperature [7]. Besides, they can inhibit [8] a growth of HAp crystals leading to a more active powder in subsequent sintering. Practically important task is to get through the above logic some widely accepted technological additives such as, for instance, polyvinyl alcohol (PVA). PVA is expected (i) to block crystal growth during synthesis, (ii) to improve a resistance of HAp suspension against sedimentation, (iii) to control an aggregation of HAp nanoparticles in the stage of drying. Therefore, the aim of this work is focused on the role of NH$_4$NO$_3$ (a by-product of the HAp precipitation) and the effect of PVA (added to the stock solution) on sintering of bioceramics.

EXPERIMENT

2.1. Powder synthesis

Powders of HAp were fabricated via conventional wet-precipitation technique by dropwise adding of (NH$_4$)$_2$HPO$_4$ solution (0.15-1.00 M) to the stock solution of Ca(NO$_3$)$_2$ (0.25 -1.67 M) with pre-adjusted pH at 60 C in presence of polyvinyl alcohol (PVA). The amount of PVA (added to Ca(NO$_3$)$_2$ solution) was set to 0.25 and 0.5 % with respect to the mass of HAp in two parallel runs. The pH of the mixture was maintained at a constant value (about 9) by addition of NH$_4$OH. The solution was vigorously stirred. After total addition of the (NH$_4$)$_2$HPO$_4$ solution, the suspension was matured during 30 min and then filtered without washing. The resulting precipitates were dried at room temperature during 48 hours. Dry powders were disaggregated by means of ball milling (3 min, acetone media, the proportion liquid:powder: balls was set to 2:1:3). Thus processed powders were passed through the sieve (Saatilene HiTech^{TM} polyester fabrics, cells of 200 µm).

2.2 Sintering

The samples (charge of about 1.5 g, without plasticiser) were pressed uniaxially under 25-50 MPa in a stainless steel die into 6x40 mm rectangular bars. Green density and density of sintered samples was determined by geometrical measurements assuming a theoretical density of 3.156 g/cm^3 for the HAp. Thermal treatment of the powder and compacted samples was done with isothermal hold during 4 hours at the temperatures 300-1200 C. The linear shrinkage was determined with dilatometer LIR-1400 (Russia) on the HAp samples pressed in the form of bars 6x4x10 mm in size. The samples were heated in air up to 1000 C at a ramp rate of 5 C/min.

2.3. Samples characterization

XRD patterns were obtained with CuKα radiation using DRON-3M (Russia). FTIR spectra of the powders were recorded on PE-1600 FTIR (Perkin Elmer, USA) in the range of 400-4000 CM^{-1} with scanning step 4 CM^{-1}. TGA of the specimens was conducted with Diamond Pyris (Perkin Elmer, USA) in air up to 1000 °C at a ramp rate 5°C/min. Microstructure of the powders and dense specimens was observed by FESEM at 5-10 kV (LEO Supra 50 VP, Carl Zeiss, Germany) and TEM at 200 kV (JEM-2000 FXII, JEOL, Japan). Composition of the samples (Ca/P ratio) was determined by SEM/EDX (INCA Energy +, Oxford Instruments, UK).

RESULTS AND DISCUSSION

According to XRD all as-synthesised powders consisted of nanocrystalline HAp (size of crystals accessed from peak broadening was in the range of 15-40 nm) with Ca/P close to 1.67. Besides the apatite phase, a significant amount of NH_4NO_3 in dry powders was revealed by FTIR (pronounced band at 1380 cm^{-1}). This amount, rulled out of TG-curves (weight change in the region 20-400°C since NH_4NO_3 decomposes at 210°C) and weight gain of as-precipitated powders against theoretical yield of the reaction, was varied from 5.87 to 27.40 % (Table 1). TEM gave a strong evidence of needle-like morphology of HAp particles, the samples with PVA demonstrating higher aspect ratio compared to the case of powder without PVA ((150...200)x(20...25) and (150...200)x(10...15) nm, respectively). The particles were seemed to be composed of primary crystals with sizes close to that determined by XRD. Another indicative feature deducing the origination of the particles from gel-like amorphous precursor was their microporosity with pore sizes less than 5 nm. The effect of PVA on microstructure of the powders consisted in formation of large aggregates detected by TEM and SEM (Fig. 1 a,b)

As one can see in Fig.1 c,d, demonstrating cleaved surfaces of compacts after thermal treatment (300°C, 4 hs), primary aggregates form bulk sintered agglomerates (in some cases more than 10 μm in size). Formation of such agglomerates was accompanied by shrinkage of the compacts in the range of 200-300°C (Fig.2), increasing with PVA content and starting concentration of salts in the stock solution (i.e., with the amount of NH4NO3 captured by the precipitate) increase. Thus, we believe that observed agglomeration of the HAp powders is a concerted action of NH4NO3 and PVA. According to FTIR and SEM (see distinct pocks in Fig.1 c,d) the former was released from the compacts at 300°C. As it followed from TG and annealing experiments PVA was burnt out at 600-700°C. Its role consisted mainly in formation of the agglomerates, i.e., gentle coarsening of the powders, which provided a close packing of the particles at rather low temperature. Hence we infer that PVA can keep high reactivity of wet-chemistry HAp powders. However, the effect of PVA on the main stage of sintering is non-monotonous one (Fig.2), revealing better sintering for the sample with intermediate content of PVA (0.25 %).

Table I Synthesis conditions and properties of powders.

Starting concentrations			Weight loss of the samples at 400°C, %	Apparent density of powders, g/cm3	Green density of compacts, % (P=50 MPa)
c(Ca^{2+}) , M	c(PO$_4^{3-}$), M	c(PVA), %			
0.25M	0.15M	0	5.87	0.28	39
0.5M	0.3M	0	13.55	0.36	40
1.0M	0.6M	0	23.99	0.42	42
1.0M	0.6M	0.25	25.39	0.56	47
1.0M	0.6M	0.5	27.40	0.67	48
1.6M	1.0M	0	27.40	0.40	46

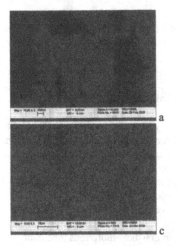
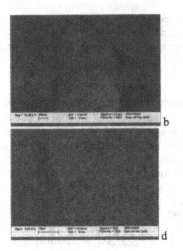

Figure1. SEM micrographs of as-synthesised powders (a) without PVA, (b) with 0.25% PVA; and treated at 300°C powders (c) without PVA, (d) with 0.25% PVA

Viable explanation of this phenomenon could be done in terms of self-organization approach. Highly dispersed powders agglomerate during molding giving rise to local densification regions due to self-organisation processes [9-12]. This implies that preliminary stages of ceramics technology can be regarded as mass-transfer process, which determine further structural evolution of the system via plastic deformation or brittle disintegration. If the amplitude of external and internal factors influencing the system in an unstable state exceeds that of noises (i.e., weak uncontrollable external perturbations and internal fluctuations), they can turn into control signals.

In our case, the powder after synthesis was agglomerated in the presence of a surfactant (PVA), which simultaneously acted as a binder. An increase in the PVA content from 0.25 to 0.5% makes the aggregates larger, stronger, and denser. Aggregates with high densities can be regarded as control signals. They provide a higher apparent density of the powder. Consequently, density of the green compact after molding also appears higher (Table I). The aggregates preserved after molding start to sinter at a lower temperature and thus densification occurs dominantly in already densified regions, which gives rise to a strong inhomogeneous framework composed of local densification regions. Shrinkage proceeds slowly, the crystals forming the framework undergo densification and rapidly loose activity towards further sintering. As a result, low-density ceramics is produced.

In contrast, in the case of a low PVA content (0.25%), the aggregates are smaller, less strong and dense; apparent density of the powder is lower; and density of the green compact after molding is respectively lower (Table 1). These aggregates do not play a role of the control signal. Although particles constituting the aggregates can loose activity during sintering, the aggregates are small and start to move upon shrinkage-induced deformations. Substantial deformations disintegrate the aggregates that are characterised by a low strength. This results in large shrinkages (Fig.2) and, consequently, high-density ceramics.

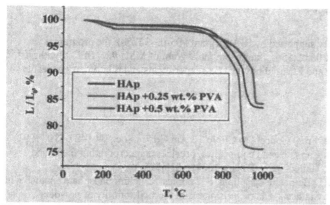

Figure 2. Linear shrinkage of the compacts (P=25 MPa) with different amount of PVA

Finally, in the absence of PVA, strong interparticle friction favours formation of local densification regions during molding. Then, the local densification regions are sintered giving rise to a strong framework, like in the case of the high PVA content (0.5%), the total shrinkage is reduced and ceramics produced appears porous. This occurs due to self-organisation, in contrast to the case of the high PVA content, where the granules were pre-prepared in the molding stage.

Reduction of sintering temperature from 1150°C according to the data [13] down to 900-950°C in our work (which is of special interest) can originate from higher sintering activity of the powder under study. Sintering activity of a powder is a qualitative characteristic depending mostly on the fabrication route. The increasing of crystal lattice imperfection and increasing of specific surface area of powder, reduction of the aggregation extent make the sintering activity of the powder higher. In our case sintering activity was enhanced due to shortening of duration of precipitation and subsequent aging of the precipitate, lowering of drying temperature and using special program of desaggregation of the powder.

CONCLUSIONS

Addition of PVA to the stock solution in the course of HAp precipitation is a promising technique to control an aggregation of HAp nanoparticles in the stages of drying and sintering. PVA acting as a surfactant in the solution and as a binder in dry powder can keep highly reactive small HAp particles within large agglomerates providing better molding of the powder and controllable densification of ceramics. Ammonium nitrate coming from the mother liquor facilitates the agglomeration of the powders due to melting at 180°C and subsequent evaporation at 210°C. However, in order to obtain high green density of the compacts the amount of NH_4NO_3 has to be diminished in comparison to that found in unwashed precipitates.

ACKNOWLEDGEMENTS

This work was partially supported by RFBR (grant #05-03-32768), the programm 'Universities of Russia'(UR 06.02.556), interdisciplinary grant of MSU #26, SEC 'Synthesis'. The assistance of N.Lyskov and R.Muydinov in TGA-experiments and dilatometry is deeply appreciated.

REFERENCES

1. L.L.Hench, "Bioceramics: From Concept to Clinic", J.Am.Ceram.Soc., 74 1487-510 (1991)
2. Ying, et al. Nanocrystalline apatites and composites, prostheses incorporating them, and method for their production, United States Patent, 2000, 6,013,591
3. S.Raynaud, E.Champion, D.Bernache-Assollant, P. Thomas "Calcium phosphate apatites with variable Ca/P atomic ratio I. Sinthesis, Characterisation and thermal stability of powders", Biomaterials 23, 1065-1072 (2002).
4. V.L.Balkevich, A.V Belyakov., T.A. Safronova "Synthesis of disaggregated fine mullite powder by meanse of chemical methods" Steklo I keramika, №5, 25-27 (1985), (in Russian).
5. Massala Ombretta and Ram Seshardi. "Synthesis Routes for Large Volumes of nanoparticles" Annu.rev.Mater.Res. 34, 41-81 (2004).
6. Satyabrata Si, Atanu Kotal, Tarun K.Mandal, Saurav Giri, Hiroyuki Nakamura, and Takao Kohara, "Size-Controlled Synthesis of Magnetite Nanoparticles in the presence of Polyelectrolytes" Chem. Mater. 16, 3489-3496 (2004).
7. A.V. Belyakov "Methods of preparation of inorganic nonmetallic nanoparticles" (MCTUR, Moscow, 2003), 80 p (in Russian).
8. Z.Amid "The influence of Polyphosphates, Phosphonates, and poly(carboxylic acids) on Crystal Growth of hydroxyapatite" Langmuir 3, 1063-1069 (1987)
9. A.V.Belyakov, A.S.Yenko, "Regions of Local Densification and their Role in Ceramics. Analysis of Fracture Surfaces as Simple Tool to Study the Local Densification Regions in Oxide Ceramics, Fracture Mechanics of Ceramics" V. 13. Crack-Microstructure Intyeraction, R-Curve Behavior, Enviromental Effect in Fracture, and Standartization. (N-Y., Boston, Dordrecht, London, Moscow: Kluwer Academic/Plenum Publishers, 2002) - 529 p. - P. 497 - 502.
10. A.V. Belyakov, "Principal Bifurcations in Firing of Compact Oxide ceramics", Glass and Ceramics 57, № 10, 345-349 (2000)
11. I. Prigogine, The End of Certitude. (Free Press, N.Y.1997).
12. A.V.Belyakov, "Synergetic and Quasichemical Approaches in Ceramic Technology" (a Review) Glass and Ceramics. 60, № 9 – 10, 274 – 279 (2003).
13. S. Raynaud, E. Champion, D. Bernache-Assollant, P. Thomas, "Calcium phosphate apatites with variable Ca/P atomic ratio II. Calcination and sintering", Biomaterials, 23, 1073-1080 (2002).

Mater. Res. Soc. Symp. Proc. Vol. 887 © 2006 Materials Research Society
0887-Q06-05

Ion Desorption by Inner-shell Excitation and Photodegradation of Poly(vinylidene fluoride)

Koji K.Okudaira, Eiichi Kobayashi[1,2], Satoshi Kera, Kazuhiko Mase[1,3], and Nobuo Ueno
Faculty of Engineering, Chiba University, Chiba, 263-8522, Japan
[1]Institute of Materials Structure Science, Tsukuba, Ibaraki, 305-0801, Japan
[2]Inoue Foundation for Science, Shibuya-ku 150-0036, Japan
[3]PRESTO, Japan Science and Technology Agency, Kawaguchi, 332-0012, Japan

ABSTRACT

Auger electron spectra (AES) and Auger electron photo-ion coincidence (AEPICO) spectra of poly(vinlylidene fluoride) (PVDF) were observed to study the mechanism of effective fluorine ion (F^+) desorption by irradiation of photons corresponding to the transition from F 1s to $\sigma(C-F)^*$. In the AES at photon energy (hv) = 690.3 eV ($\sigma(C-F)^* \leftarrow$ F 1s), the spectator-Auger shift is about 3 eV. At hv = 690.3 eV the contribution of spectator Auger electron to the observed Auger spectrum is large. In the F^+ AEPICO spectra at hv = 690.3 eV an intense peak appears. The peak positions of the F^+ AEPICO spectra are in agreement with those of the Auger spectra. These results indicate that the mechanism for effective F^+ ion desorption induced by the transition from F1s to $\sigma(C-F)^*$ occurs through the spectator-Auger processes. The efficiency of ion desorption depends on the electronic structure of the spectator-Auger final state.

INTRODUCTION

Vinylidene fluoride and trifluoroethylene copolymer is a well-known ferroelectric material which is a promising material for electrical devices in future, such as high density memories, and piezoelectric sensors or actuators in sub-millimeter size.[1]   In the field of bioscience and medicine, furthermore, fluorocarbon polymers such as poly (vinylidene fluoride) (PVDF, $(-CF_2-CH_2-)_n$) may be one of the most suitable materials for manufacturing microparts due to the chemical inertness and the thermal stability.   However, it is difficult to apply PVDF to the micro-machining technology since there are few solvents to dissolve the polymers and heating over their melting points does not result in enough fluidity for modeling.

Recently, it has been reported that the photodegradation and surface modification of PVDF film occur easily under electron, ion and x-ray irradiation.[2,3]   These reports show that PVDF is a polymer in which scissions occur primarily between the chain backbone carbons and the side substitutions, leading to the formation of double bonds in the main chain [4]   Furthermore, by using a synchrotron radiation, it is shown that the effective fluorine ion (F^+) desorption for PVDF is induced by the irradiation of photon corresponding to the transition from F 1s to $\sigma(C\text{-}F)^*$.[5] It indicates that C-F bond scission effectively occurs following the transition from C 1s to $\sigma(C\text{-}F)^*$.

The analysis of desorbed ion by inner-shell excitation is one of useful methods for clarification of the relationship between the mechanism of bond scission and the electronic configuration of excited states.   For not only simple molecules but also polymers with large molecular weight and complex structure, there have been extensive works on the photon-stimulated ion desorption (PSID) by inner-shell excitation.[6-8]   For PSID, an Auger-stimulated ion desorption (ASID) mechanism has been proposed.[9]   Coincidence measurements of energy-selected electrons and ions have been developed as a method for ion desorption studies.   Among coincidence spectroscopies, the Auger-electron photo-ion coincidence (AEPICO) method has become one of the most powerful tool because it can be used to measure ion desorption yields for a selected core excitation or a selected subsequent Auger transition.[10]

EXPERIMENTAL DETAILS

PVDF was supplied by Kureha Chemical Industry Corporation Ltd. (mean value of molecular weight (Mw) is about 5000).   The PVDF thin film was evaporated on copper plates with thickness of 130 Å.

Experiments were performed at the beamline 8A at the Photon Factory, Institute of Materials Structure Science.   AEPICO spectra were measured by using the EICO apparatus, which is composed of a coaxially symmetric mirror electron energy analyzer [11] and a polar-angle-resolved compact time-of-flight ion mass spectrometer with four concentric anodes.[12]   Total electron yield (TIY) and partial electron yield (PEY), Auger, and AEPICO spectra were observed at the incidence angle of the photons of 84° (grazing incidence). TIY and PEY were normalized to the incident photon flux recorded as the photocurrent at the photon-flux monitor consisting of a gold-evaporated mesh.   All measurements were performed at room temperature.

RESULTS AND DISCUSSION

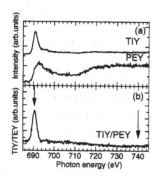

Figure 1 (a) Total ion yield (TIY) and partial electron yield (PEY) with electron kinetic energy of 640 eV of PVDF near the fluorine K-edge region. (b) The spectrum of TIY divided by PEY. The arrows indicate the photon energies for AEPICO TOF spectra in Fig. 2 and AES in Fig. 3.

Figure. 1 (a) shows total ion yield (TIY) and partial electron yield (PEY) spectra of PVDF near the fluorine (F) K-edge region. In the TIY spectrum an intense peak appears at photon energy (hv) = 690.3 eV. On the other hand, in the PEY spectrum a broad peak consisting of two peaks at hv = 690 eV and 692eV is shown. Peaks at hv = 690 eV and 692 eV are ascribed to the transition from F 1s to $\sigma(C\text{-}F)^*$ and $\sigma(C\text{-}C)^*$, respectively, by taking account into the peak assignment of NEXAFS near the F K-edge for polytetrafluoroethylene [13]. In Fig. 1 (b) the spectrum of TIY divided by PEY (TIY/PEY) exhibits an intense peak at hv = 690.3 eV. The energy position of the peak in TIY/PEY spectrum agrees with that of the transition of $\sigma(C\text{-}F)^*$ $\leftarrow$ F 1s. It indicates that the effective ion desorption occurs by the irradiation of photons, corresponding to the transition of $\sigma(C\text{-}F)^*$ $\leftarrow$ F 1s. As we reported previously, the ions desorbed from PVDF due to the irradiation of photons near F 1s region are F^+ and H^+ [5]

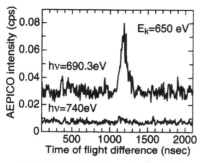

Figure 2 Typical AEPICO spectra of PVDF at hv = 690.3 eV and 740 eV in coincidence with electron emission at E_k = 650 eV.

Among these ions, the partial ion yield spectrum of F^+ near F 1s region which shows an intense peak at the transition of $\sigma(C-F)^*$ ← F 1s is similar to the TIY spectrum shown in Fig.1 (a). Therefore, the major ion species in the TIY is considered to be F^+. The highly F^+ desorption at hv = 690.3 eV indicates that the C-F bond scission in PVDF molecule occurs effectively by the irradiation of photons.

Figure 2 shows typical Auger electron photo-ion coincidence (AEPICO) time-of-flight (TOF) spectra of PVDF in coincidence with electron emission at a kinetic energy (E_k) of 650 eV upon excitation at hv = 690.3 eV ($\sigma(C-F)^*$ ← F 1s) and 740 eV (above F 1s ionization energy). A peak was observed at the TOF difference of about 1200 ns at hv = 690.3eV. At hv = 740 eV the peak intensity is much weaker than that at hv = 690.3 eV. It indicates that the ion desorption corresponding to the AEPICO peak at the TOF difference of 1200 ns at hv = 690.3 eV is much higher than that at hv = 740 eV. This hv dependence of the AEPICO peak intensity is similar to that of F^+ ion yield in the TIY spectrum shown in Fig.1 (a). Therefore, the AEPICO peak at the TOF difference of about 1200 ns could be assigned to F^+ ion.

Figure 3 shows Auger electron spectra (AES) of PVDF at hv = 690.3 eV ($\sigma(C-F)^*$ ← F 1s) and 740 eV (above F 1s ionization energy). In the spectra at hv = 690.3 eV an intense peak and a broader peak appear at E_k about 655 eV and 610 eV, respectively. In the F KLL AES of fluorides such as NaF and MgF_2, the AES peak at E_k about 652 eV was assigned to $KL_{23}L_{23}$.[14] From this, in the AES at hv = 690.3 eV, the intense peak at E_k about 655 eV could be assigned to the spectator Auger final state, where two holes are created at $\sigma(C-F(2p))$ bonding state and one excited electron remains in $\sigma(C-F)^*$ antibonding state. The whole structure of AES at hv = 740 eV is much broader than that at hv = 690.3 eV. In detail, however, the energy position of the intense Auger peak at hv = 690.3 eV is shifted to larger E_k of about 3 eV, as compared with that at hv = 740 eV. In general, the kinetic energy of the spectator-Auger electron is larger by a few eV than that of the corresponding normal-Auger electron due to the Coulomb interaction

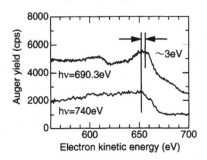

Figure 3 Auger electron spectra of PVDF at hv = 690.3 eV and 740 eV.

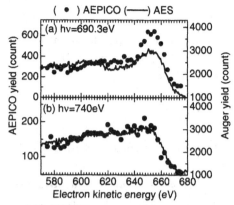

Figure 4 F⁺ AEPICO spectra and Auger electron spectra (AES) of PVDF at hv = 690.3eV and 740 eV.

between the Auger electron and the electron excited to the unoccupied state.[15]   It indicates that the excited electron in σ(C-F)* is localized and the interaction between excited electron in σ(C-F)* and the Auger electron is strong.   At hv = 690.3 eV the contribution of spectator Auger electron to the observed Auger peak at E_k = 655 eV is considered to be large.

Figures 4 (a) and (b) show the F⁺ AEPICO spectra of PVDF at hv = 690.3 eV (σ(C-F)* ← F 1s) and 740 eV (above F 1s ionization energy), respectively.   For the comparison, the AES are also represented.   The peak positions of the AEPICO spectrum at hv = 690.3 eV are in agreement with those of the AES.   At hv = 690.3 eV, a strong enhancement of F⁺ AEPICO yield appears at E_k = 650-660 eV corresponding to the Auger final state ((valence)σ(C-F(2p)))⁻² σ(C-F)*¹).   At E_k about 610 eV corresponding to the broad Auger peak, the F⁺ AEPICO yield does not enhance.   On the other hand, at hv = 740 eV, where the normal-Auger process occurs mainly, the APICO yield spectrum agrees with the Auger electron yield and the enhancement of F⁺ AEPICO yield does not appear.

From these results, a possible explanation of selective F⁺ ion desorption at hv = 690.3 eV is that the desorption occurs through the spectator-Auger processes. The electronic structure of the Auger final state as well as the excited state by the core excitation affect the efficiency of bond scission.   At the transition from F 1s to σ(C-F)*, C-F bond breaking by spectator-Auger process is more effective than that by the normal-Auger process since the final states of spectator-Auger process C-F bonds are more repulsive than those at the normal Auger final states.

ACKNOWLEDGEMENTS

This research was supported in part by the Grant-in-Aid for Scientific Research (16560021) and 21st Century COE Program (Frontiers of Super-Functionality Organic Devices) from the Ministry of Education, Science, Sports and Culture.

REFERENCES

1 A. V. Bune, V. M. Fridkin, S. Ducharme, L. M. Blinov, S. P. Palto, A. V. Sorokin, S. G. Yudin, A.Zlatkin, *Nature*, **391**, 874 (1998).

2. A. Le Moel, J. P. Duraud, and E. Balanzat, *Nucl. Instrum. Methods. Phys. Res.* **B18**, 59 (1989).

3. M. D. Duca, C. L. Plosceanu, and T. Pop, *J. Appl. Polym. Sci.*, **67**, 2125 (1995).

4. E. Morikawa, J. Choi, H. Manohara, H.Ishii, K. Seki, K. K. Okudaira, and N.Ueno, *J. Appl. Phys.* **87**, 4010 (2000).

5. K. K. Okudaira, H. Yamane, K. Ito, M. Imamura, S. Hasegawa and N. Ueno, *Surf. Rev. and Lett.* **9**, 335 (2002).

6. A. P. Hitchcook, P.Lablanquie, P. Morin, E. Lizon, A. Lugrin, M. Simon, P. Thiry, and I. Nenner, *Phy. Rev.*, **A 37**, 2448(1988).

7. D. Coulman, A. Puschmann, W. Wurth, H. -P. Steinrück, and D. Menzel, *Chem. Phys. Lett.*, **148**, 371 (1988).

8. M. C. K. Tinone, K. Tanaka, J. Maruyama, N. Ueno, M. Imamura, and N. Matsubayashi, *J. Chem. Phys.*, **100**, 5988 (1994).

9. D. E. Ramaker, in : W. Brenig, D. Menzel (Eds.), Desorption Induced by Electronic Transitions, DIET-II, Springer Series in Surface Sciences, Springer, Berlin, **14**, 10 (1985).

10. K. Mase, M. Nagasono, S. Tanaka, T. Sekitani, and S. Nagaoka, *Fizika Nizkikh Temperatur*, **29**, 321 (2003). (*Low Temp. Phy.* **29**, 243-258(2003)).

11 K. Isari, H. Yoshida, T. Gejo, E. Kobayashi, K. Mase, S. Nagaoka, and K. Tanaka, *J. Vac. Soc. Jpn.*, **46**, 377 (2003) .

12 E. Kobayashi, K. Isari, M. Mori, A. Nambu, and K. Mase, *J. Vac. Soc. Jpn.* **47** 14 (2004)

13. T. Ohta, K. Seki, T. Yokoyama, I. Morisada, and K. Edamatsu, *Phys. Scri* **41**, 152 (1990).

14. K. Kover, M. Uda, I. Csermy, J. Toth, J. Vegh, D. Darge, K. Ogasawara, and H. Adachi, *J. Vac. Sci. Tech.* **A19** 1143 (2001) .

15. T. A. Sasaki, Y. Baba. K.Yoshii, and H Yamamoto, *J. Phys. Condensed Matter*, **7**, 463 (1995).

Mater. Res. Soc. Symp. Proc. Vol. 887 © 2006 Materials Research Society   0887-Q06-08

Degradation Phenomenon in Semiconductive Nanoamorphous Structures from Molybdenum Oxide.

Markosyan A., Kamanchadgyan S., Malkhasyan R.
Scientific Production Enterprise "ATOM", Tevosyan st. 3/1 Yerevan 375076, Republic of Armenia, email: samar1@freenet.am

ABSTRACT

The condition of degradation of electrical properties of samples prepared from powder of nanoamorphous semiconductive oxide in the course of time is discovered and studied when investigating electro-physical properties of the given materials for the first time synthesized at Scientific Production Enterprise "ATOM". It has been shown that degradation phenomenon is also present in amorphous materials, which relies on the configuration of the pores of the sample.

The resistance of sample in the open air changes over a period of several months. However, it decreases and is comparatively fast stabilized when placed in vacuum. The absorption and de-absorption of different gases have diffusion nature and highly depend on environment temperature. Water steam in atmosphere has been determined to be the main cause of degradation.

INTRODUCTION

It is well known that at the beginning the nano-crystal materials played important role in evolution of nanotechnology [1]. Nanoamorphous material investigation gained significant interest in recent years [2, 3]. At the time being one of the major interests for scientists and engineers is to obtain highly dispersive and sensitive materials which are nanoamorphous substances. Amorphous materials possess bigger specific surface area and have additional activity due to presence of particles with surplus internal energy.

Nevertheless appliance of nanoamorphous materials in technology is still limited, because of high sensitivity of the latter to environmental conditions. For example, a degradation phenomenon can be observed on samples prepared from powder of Nanoamorphous Molybdenum Oxide ($MoO^{\bullet}_2$), which shows up in daily increase of resistance. The increase in resistance in one order has been recorded at usual atmospheric conditions.

Degradation phenomenon has already been investigated [4] and is explained by presence of dislocations in material. However, the causes of degradation described in the present paper for a sample prepared from molybdenum oxide are completely different as the pure amorphous material is free of dislocation phenomenon.

Here results of investigation of degradation phenomenon performed on $MoO^{\bullet}_2$ samples [3] are presented.

It is shown that $MoO^{\bullet}_2$ samples expose degradation phenomenon, which is, however caused by configuration of pores and types of gases absorbed by these pores.

EXPERIMENT

The degradation phenomenon has been observed and investigated on two-contact samples (with size 14.5×6.5×1.5 mm) prepared from powder of Nanoamorphous Molybdenum Oxide (MoO$^•_2$) via mechanical pressing. This oxide (grain size 10÷15 nm) has been made from nanoamorphous molybdenum Mo (grain size 3÷4.5 nm) under strongly controllable conditions.

In our work the change in the resistance of the sample over time has been taken as a criterion of degradation.

The samples have been placed in a vacuum camera where they passed a thermo-vacuum treatment. Firstly, the pressure of the camera was brought to 10^{-3} Pa and samples were gradually heated till 200 °C. After the samples have been cleared from previously absorbed gases (which has been confirmed by stabilization of the samples' minimal resistance), different gases were sequentially inserted into the camera, gradually increasing the pressure from 10^{-3} Pa to 10^5 Pa. Before inserting the next gas the pores of the samples have been cleared from previous gas using the above mentioned thermo-vacuum treatment steps. The changes of resistance of the samples as well as the pressure, temperature and humidity have been measured over time for each gas.

RESULTS AND DISCUSSION

Investigations showed that in open air the resistances of MoO$^•_2$ samples have been continuously increasing for a period of several months from the moment of production of the samples.

It was possible to stop the increase of resistances (degradation) of the samples by lowering the environment pressure. This can be seen in figure 1.

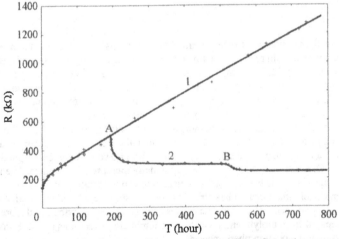

Figure 1. The dependence of resistances of the samples over time. Curve 1 presents the change in resistance of a sample under normal atmospheric conditions. Curve 2 presents the change in resistance of a sample which has been kept in the normal conditions till point A, then from point A to point B it was kept under 10 Pa air pressure and 1 Pa pressure from point B.

Curve 1 presents the change in resistance of a sample under normal atmospheric conditions. Curve 2 presents a similar behavior of another sample which has been kept in the same normal conditions till point A, then from point A to point B it was kept under 10 Pa air pressure and 1 Pa pressure from point B.

From these curves it can be concluded that it is the environment – atmosphere, which results in resistance increase.

The relative changes of resistances of the samples depending on changes of temperature, environment pressure as well as changes of composition of the environment with time have been studied. The dependence of resistance changes on changes of the environment composition shows the degree of sensitivity of the material to various gases. This has been studied for nitrogen (N_2), argon (Ar), hydrogen (H_2), methane (CH_4), oxygen (O_2) and water steam (H_2O). The dependence of resistance on temperature has been studied under various air pressure conditions (10^5 Pa, 1 Pa and 10^{-3} Pa). The results of 1 hour period are shown in the table I. The $\Delta R/R$ in α coefficient, is the resistance relative change in the given temperature interval Δt. In the table I, the data shown are for a temperature interval from 20 ^{0}C to 70 ^{0}C.

Table I. Temperature coefficient of resistance for different air pressure values.

Air Pressure (Pa)	Temperature Coefficient of Resistance ($\alpha = \Delta R/(R \times \Delta t)$) (1/^{0}C)
10^5	-0.6×10^{-2}
1	-2×10^{-2}
10^{-3}	-2.73×10^{-2}

As evident from the table, the α coefficient is negative. The latter is typical for semi-conductive materials.

The investigation of resistance changes at a high temperature interval can be done only in vacuum to escape the influence of the environment on the material.

The α coefficient has been determined for pressure of 10^{-3} Pa and for temperature from 0 ^{0}C to 200 ^{0}C. In that interval the temperature α coefficient is not constant and is laid within the range from $-0.9 \times 10^{-2}/^0$C to $-0.5 \times 10^{-2}/^0$C.

It is significant to note that during the investigation of amorphous materials in high temperature conditions, special care must be taken to avoid crystallization of the material even in vacuum (especially during protracted experiment).

The investigation of samples' resistance dependence on environment pressure has been performed at 20 ^{0}C temperature. The resistance change coefficient calculated for pressure range from 10^5 Pa to 10 Pa and 10 Pa to 1 Pa, is shown in table II. The $\Delta R/R$ in β coefficient, is the resistance relative change in the given pressure interval ΔP.

Table II. Pressure coefficient of resistance for different air pressure ranges.

Air Pressure Range (Pa)	Pressure Coefficient of Resistance ($\beta = \Delta R/(R \times \Delta P)$) (1/Pa) in 20 ^{0}C
$10^5 \div 10$	0.8×10^{-7}
$10 \div 1$	1.8×10^{-4}

Heating of the sample sharply accelerates the resistance change at pressure variations near atmospheric. As appears from the above said and from the values given in the table II, the sample's resistance change is determined by the diffusion phenomenon, which depends on pressure (the free path of the absorbed gas's molecules) and temperature (the speed of the absorbed gas's molecules). The diffusion coefficient is as larger as the pressure is low and the temperature is high.

The resistance change depending on pressure and temperature simultaneous changes is presented by temperature and pressure coefficient of resistance $\gamma=\Delta R/(Rx\Delta tx\Delta P)$. The $\Delta R/R$ in γ coefficient, is the resistance relative change in the given temperature Δt and pressure ΔP intervals. It has a value $0.2x10^{-6}/(^0CxPa)$ for temperature range from 15 ^{0}C to 60 ^{0}C and pressure range from 10^5 Pa to 10^{-3} Pa.

The last value of γ is very close to the product of the maximum available values of coefficients α and β as $\Delta R/R \approx 1$. It again justifies the fact that it is the diffusion phenomenon of the gases absorbed by pores of the sample which influences the change in resistance. In fact the samples' resistance changes mainly depend on the type of the absorbed gas, pressure and temperature which has also been mentioned [5].

The sample's resistance irreversibly changes during the simultaneous change of temperature and pressure. This is caused by purification of pores of the sample (figure 2 A and B points). In figure 2 the angular coefficient around point A almost does not change even after irreversible change of the resistance (the de-absorption phenomenon characteristics remain). However, at temperature range 60÷65 ^{0}C we observe a stepwise change of the angular coefficient around point B. The latter is a result of change of de-absorption characteristics. Till the point B the physical de-absorption dominates, and chemical de-absorption becomes dominant after the point B.

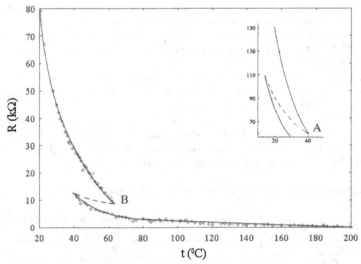

Figure 2. The change of sample's resistance during termo-vacuum processing.

The influence of different gases such as nitrogen (N_2), argon (Ar), hydrogen (H_2), methane (CH_4), oxygen (O_2) and water steam (H_2O), on the samples has been studied under different pressures. Before studying the influence of any of the above gases the pores of the samples had been purified by thermo-vacuum method. The degree of pureness has been confirmed by stabilization of the samples' minimal resistance. At this minimal resistance (around 130 Ohm) the pressure of the environment has been changed in the range from 10^{-3} Pa to 10^5 Pa at a constant temperature for all the mentioned gases except water steam, for which the maximum pressure has been set to 1.2×10^3 Pa. Table III shows the influences of the gases on the samples' resistances (the sensitivity of MoO^*_2), calculated for specific time unit.

Table III. The sensitivity of MoO^*_2 to different types of gases.

	Pressure (1/Pa)					
	N_2	Ar	H_2	CH_4	O_2	H_2O
$\Delta R/(R \times \Delta P)$	0.1×10^{-6}	0.3×10^{-6}	0.3×10^{-6}	0.7×10^{-6}	1.7×10^{-6}	3298×10^{-6}

As we can see from the Table III, water steam has the biggest influence. We could suppose that the influence of the steam was accompanied by a chemical interaction creating hydrates and other chemical compounds. But after removing the steam, the resistance of the sample changed to its initial value which accounts for the fact that no irreversible chemical interactions occur.

We should also mention that the stabilized minimal resistance values of the samples, achieved by means of thermo-vacuum processing after the influence of nitrogen (N_2), argon (Ar) and hydrogen (H_2) gases, at the same temperature, was nearly twice smaller. It speaks for the fact that these gases additionally clear the pores of the samples.

After clearing the pores by thermo-vacuum method as well as passing argon gas for additional purification, the minimal resistances of the samples stabilized around 70 Ohm value. Afterwards, the influence of water steam on resistance has been investigated for 140 hours for these samples (figure 3). The temperature and pressure of the environment had fixed values 10^0C and 1.2×10^3 Pa respectively.

From figure 3, it can be seen that during the influence of the steam, the timing coefficient of the resistance $\Delta R/(R \times \Delta T)$ in the course of time changes in the range from 5.7×10^{-2}/hour to 0.2×10^{-2}/hour. According to figure 1 the timing coefficient for the sample left in the open air has a value 0.38×10^{-2}/hour. The comparison of these two values, counts for the assumption that the degradation of the samples are mainly caused by water steam present in atmosphere.

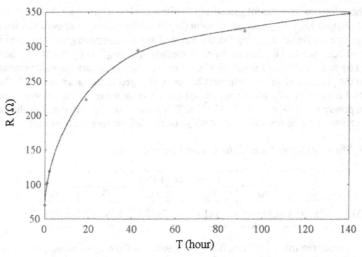

Figure 3. Dependence of sample's resistance on the (of pressure of 1.2×10^3 Pa) interaction time with water steam.

CONCLUSION

In this paper the degradation phenomenon has been investigated on samples prepared from nanoamorphous molybdenum oxide MoO_2 powder. It has been shown that degradation phenomenon is also present in amorphous materials, which relies on the configuration of the pores of the sample. The absorption and de-absorption of different gases have diffusion nature and highly depend on environment temperature. Water steam in atmosphere has been determined to be the main cause of degradation. Water steam under the pressure of 1.2×10^3 Pa have 3 times bigger influence on a resistance than the influence of oxygen with pressure of 10^5 Pa under the same conditions.

REFERENCES

1. Nanophase Materials, Sinthesis – Propeties – Applications. Edited by Hajipanjis K. and Siegel R. Kluwer Acad. Publ. p. 808, 1994.
2. Malkhasyan P.T. et al., Patent of the USSR N 2041959, 11.11.1990. "Method of direct reconstruction of oxides and making of amorphous materials".
3. Markosyan A., Grigorya S. and Krmoyan R., "Semiconductor Micro- and Nanoelectronics", proceedings of the fifth international conference, p. 94-97, Agveran, 2005.
4. Ha S. and Bergman J. P. MRS Bulletinr, v. 30, 305-307, 2005.
5. Shimizu Y. and Egashira M. MRS Bulletin, v. 24, 18-24, 1999.

Mater. Res. Soc. Symp. Proc. Vol. 887 © 2006 Materials Research Society 0887-Q08-02.1

Modeling of Coarsening Processes in Patterned Systems

M. H. Jhon, A. M. Glaeser and D. C. Chrzan[1]
Department of Materials Science and Engineering,
University of California, Berkeley, CA 94720, U.S.A.
Materials Sciences Division,
Ernest Orlando Lawrence Berkeley National Laboratory, Berkeley, CA, 94720, U.S.A.

ABSTRACT

Although particle coarsening has been studied for over a century, it remains an active area of materials science research. The current work presents a theoretical analysis of the degradation of regular arrays of spherical particles through diffusional interaction. In order to understand the onset of coarsening, a linear stability analysis is performed on a simple square lattice of particles. It is predicted that particles will dissolve in a spatially ordered manner. The active transport mechanism plays a strong role in the selection of the coherent growth modes.

INTRODUCTION

The recent study of nanostructures has renewed interest in the stability of material features. At elevated temperatures, microstructures may evolve in order to minimize the total surface energy of a system. In particular, larger particles tend to grow at the expense of smaller ones. In many different materials systems, particles coarsen through an evaporation-condensation mechanism. This phenomenon has since become known as Ostwald ripening. Quantitative theories explaining this process were first developed by Lifshitz and Slyosov and independently by Wagner using mean-field techniques [1, 2]. This so-called LSW approach concentrates on understanding the asymptotic limit, in which the particle size distribution reaches a self-similar shape.

This mean field analysis is limited: it does not take into account the local environment about each particle. Modern theories of coarsening have improved on this aspect of LSW theory using various statistical averaging methods [3–5]. Another shortcoming comes from the difficulty reaching the asymptotic regime [6]. Instead, transient effects may dominate the coarsening process. It has further been shown that in certain system geometries, the coarsening process is only approximately self-similar [7].

The present work addresses these two concerns by studying the onset of coarsening for a known particle size distribution. In this study, a model two-component system is considered in which one phase is dispersed in a matrix phase. Elastic effects are ignored. A linear stability analysis is applied to a regular, 2-dimensional array of spherical particles. This model system is chosen because of the existence of lithographic techniques to pattern regular arrays of features at a grain-boundary [8]. Further, examining regular lattices may allow us to better understand materials processes in which spatial patterns persist during coarsening, such as spinodal decomposition [9] and Liesegang patterning [10].

---

[1] dcchrzan@socrates.berkeley.edu

THEORY

In order to analyze the coarsening of the particle lattice, it is necessary to solve the microscopic diffusion problem. There are a variety of approximations that may be made to simplify the diffusion problem for particle growth and dissolution, as reviewed by Aaron and Kotler [11]. This study employs the invariant field approximation in which the concentration obeys the steady state diffusion equation, Eq. 1. The concentration c is in in general a function of time t and position $\mathbf{x}$. This approximation is justified under the condition of small driving forces

$$\nabla^2 c(\mathbf{x}, t) = 0. \tag{1}$$

The equilibrium solute concentration in the matrix adjoining each particle c_i is enhanced by the curvature of the precipitate, as described by the Thomson-Freundlich equation

$$c_i = c_\infty \exp \frac{l_c}{R_i}, \tag{2}$$

where c_∞ is the equilibrium concentration at a flat surface, l_c is defined to be the capillary length of the system, and R_i is the radius of the ith particle. If a distribution of particles are not uniformly sized, a concentration gradient is present, and mass transport is expected to occur. The kinetic details of the process depend on the geometry of the diffusion paths and the rate limiting step of coarsening.

In general, particles at a grain-boundary exchange mass through both bulk and grain-boundary diffusion. The diffusion coefficient of the grain-boundary will be greater than that of the bulk. However, because the width of the grain-boundary is small, at high temperatures the effects of bulk diffusion can dominate. This diffusion problem is quite complicated to solve analytically, but may be approximately solved [12]. The present approach instead considers the geometry in two limits, (i) where bulk diffusion dominates and (ii) where grain-boundary diffusion dominates.

For both cases it is possible to pose the multi-particle diffusion problem in terms of a collection of point sources and sinks. This formulation of the coarsening problem was first examined by Weins and Cahn [13] and is valid only in the dilute coverage limit. This model has been used subsequently in several large scale simulations of systems coarsening under 3-dimensional diffusion [14, 15] as well as 2-dimensional diffusion [4].

For convenience, all lengths are chosen to be scaled by l_c and time scaled by $t_c = l_c^2/(Dc_\infty\Omega)$, where Ω is the molar volume of the minority component and D is the diffusion coefficient. Additionally, a non-dimensional concentration may be chosen to be $\theta = (c - c_\infty)/c_\infty$. Using these variables, it is possible to solve Eqs. 1 and 2 in the dilute limit to find N growth rate equations that describe the time dependence of the particle radii. If bulk diffusion mediated coarsening is assumed, these rate equations for bulk diffusion mediated coarsening can be written in the form:

$$\frac{1}{r_i} = K - \frac{B_i}{r_i} - \sum_{j \neq i} \frac{B_j}{r_{ij}}, \tag{3}$$

where r_i is the reduced radius of the ith sphere, r_{ij} is the reduced distance between the ith and jth particle, K is a constant set by mass conservation of the system and the B_i terms are time dependent constants. By applying mass conservation to each particle, it is possible to relate the rate of change of the size of the ith particle to B_i

$$B_i = r_i^2 \frac{dr_i}{dt}. \tag{4}$$

For grain-boundary diffusion mediated coarsening, an equivalent expression is found

$$\frac{1}{r_i} = K + B_i \log r_i + \sum_{j \neq i} B_j \log r_{ij}.$$ (5)

Linear Stability Analysis

Consider an array of N spherical particles, where the reduced radius of the ith sphere is r_i and the reduced lattice constant is a. An unstable equilibrium of this system exists when all particles have the same radius, r. Under the dilute coverage approximation, $r/a \ll 1$. The linear stability analysis identifies the spatial modulations that diverge fastest from the fixed point. To this end, an expansion is taken about the fixed point using a spatially modulated perturbation, $r_i = r + A e^{i\mathbf{k} \cdot \mathbf{x}_i} e^{\alpha t}$. A is taken to be a small number. The amplification factor α has the physical interpretation of being the rate at which a particular mode grows. The wavevector $\mathbf{k}$ determines the spatial periodicity of the growth mode.

RESULTS

If a preferred coherent growth mode exists, it will correspond to a wavevector $\mathbf{k}$ which maximizes α. In this section, the amplification factor α is numerically evaluated for the two diffusion geometries.

Bulk diffusion controlled limit

By performing a linear stability analysis on Eq. 3, an expression for the amplification factor in the limit of bulk diffusion controlled coarsening is found,

$$\alpha = \frac{1}{r^2 \left(1 + r\sum_{j \neq i} \frac{1}{R_{ij}} e^{i\mathbf{k} \cdot (\mathbf{x}_j - \mathbf{x}_i)}\right)} \equiv \frac{1}{r^2 (1 + (r/a)\lambda(\mathbf{k}))}.$$ (6)

This expression implies that the fastest growing modes correspond to the wavevectors which minimize the lattice sum $\lambda(\mathbf{k})$. Sums similar to $\lambda(\mathbf{k})$ appear in many areas of applied mathematics, and have been considered in the context of the thermodynamics of 3-d crystals [16]. They may be approached by applying the Ewald transformation formula and the method of θ-transforms. This technique was later applied to 2-d lattices by Lee and Bagchi [17]. Following their method, it is possible to calculate and plot $\lambda(\mathbf{k})$ along the high symmetry directions of the Brillouin zone, as depicted in Fig. 1. The fastest growing growth mode corresponds to the global minimum in $\lambda(\mathbf{k})$, which lies on the X symmetry point. A schematic of this growth mode is shown in Fig. 2.

The calculation reveals that the difference between modes is small, as λ is of order r/a. The linear stability analysis does not conclusively establish if these coherent growth modes will appear in practice. It will be necessary to perform direct numerical integration of the rate equations [18].

Grain-boundary diffusion controlled limit

Due to the logarithmic divergence present in Eq. 5, the grain-boundary diffusion controlled limit of coarsening is more difficult to analyze. In the present work, an approximation that takes

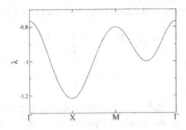

Figure 1: Plot of λ for a square lattice along high symmetry directions along the Brillouin zone.

into account the effect of nearest neighbors only is applied. This approach has been used by several authors in both the interface controlled [19] and diffusion controlled [20, 21] limits. Using the method of Zheng and Bigot [21], an expression for the flux between a pair of isolated, 3d particles can be found by solving Eq. 5 for $N = 2$. Under this limit, an expression for α may be found to be:

$$\alpha = \frac{a}{\pi r^2 \log(a^2/r^2)} \left[(1 - \cos k_x a) + (1 - \cos k_y a) \right]. \tag{7}$$

From this form, it is apparent that the maximum of α occurs at the M symmetry point. Again, the difference between modes is small. However, unlike the 3d-diffusion calculation, the amplification factor depends on the capillary length of the system. This suggests that coherent growth modes will be favored for smaller values of r.

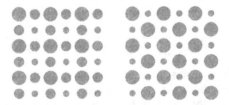

Figure 2: Representations of coherent growth modes. Larger circles represent growing particles, while smaller circles represent shrinking ones. The left panel depicts the X point instability, while the right panel depicts the the instability at M.

SUMMARY

A linear stability analysis is performed on the rate equations determining the growth of a square array of particles. For 2d diffusion, it is necessary to utilize a nearest neighbor approximation. It is found that different coherent growth modes occur in the bulk and grain-boundary controlled regimes. However, for both geometries, it is found that the difference between growth modes is small. This implies that it is necessary to numerically integrate the rate equations to verify the existence of coherent growth modes.

ACKNOWLEDGEMENTS

MHJ gratefully acknowledges support through an Intel Foundation Fellowship. This work was sponsored in part by the Director, Office of Science, Office of Basic Energy Science, Division of Materials Science and Engineering, of the US Department of Energy under Contract No. DE-AC02-05CH11231.

REFERENCES

1. I. M. Lifshitz and V. V. Slyosov, J. Phys. Chem. Solids **19**, 35 (1956).

2. C. Wagner, Z. Elektrochem **65**, 581 (1961).

3. A. D. Brailsford and P. Wynblatt, Act Metall. **27**, 489 (1979).

4. J. H. Yao, K. R. Elder, H. Guo, and M. Grant, Phys. Rev. B **47**, 14110 (1993).

5. J. A. Marquess and J. Ross, J. Chem. Phys. **80**, 536 (1984).

6. V. A. Snyder, J. Alkemper, and P. W. Voorhees, Acta Mat. **49**, 699 (2001).

7. J. Viñals and W. W. Mullins, J. Appl. Phys. **83**, 621 (1998).

8. J. Rodel and A. M. Glaeser, Materials Lett. **6**, 351 (1988).

9. J. W. Cahn, Acta Met. **14**, 1685 (1966).

10. S. C. Muller and J. Ross, J. Phys. Chem. A **107**, 7997 (2003).

11. H. B. Aaron and G. R. Kotler, Metall. Trans. **2**, 393 (1971).

12. J. J. Hoyt, Acta Met. **39**, 2091 (1991).

13. J. J. Weins and J. W. Cahn, in *Sintering and Related Phenomena*, edited by G. C. Kuczynski (Plenum, New York, 1973), p. 151.

14. P. W. Voorhees and M. E. Glicksman, Acta metall. **32**, 2001 (1984).

15. P. W. Voorhees, J. Stat. Phys. **38**, 231 (1985).

16. M. Born and M. Bradburn, Proc. of the Cambridge Phys. Soc. **39**, 104 (1943).

17. M. H. Lee and A. Bagchi, Phys. Rev. B **22**, 2645 (1980).

18. S. Reiss and K. H. Heinig, Nucl. Instr. and Meth. in Phys. Res. B **84**, 229 (1994).

19. W. Theis, N. C. Bartelt, and R. M. Tromp, Phys. Rev. Lett. **75**, 3328 (1995).

20. K. Morgenstern, G. Rosenfeld, and G. Comsa, Surf. Sci. **441**, 289 (1999).

21. X. Zheng and B. Bigot, J. Phys. II (France) **4**, 743 (1994).

Mater. Res. Soc. Symp. Proc. Vol. 887 © 2006 Materials Research Society   0887-Q09-02

Degradation of Copper / Silver Alloy Thin Films Induced by Annealing

S. T. Snyder and A.H. King
School of Materials Engineering
Purdue University, West Lafayette, IN 47907-2044

ABSTRACT

Copper / silver alloy thin films form with a fine, polycrystalline, metastable crystal structure. The expected effects of annealing include grain growth, transformation into the two stable phases, coarsening of the phases, texture formation, and the formation and growth of pinholes or voids. Copper/silver alloy films were deposited on single crystal sodium chloride substrates, *via* pulsed laser deposition ablation of an alloy target, of the eutectic composition. Free-standing films of 20-30 nm thickness were studied as-deposited and after annealing on copper TEM grids at 100°C for various times. Although several of the expected degradation processes involve short-range diffusion – essentially single atomic jumps – these were not observed, while other, longer-range diffusion effects were clearly identifiable. In particular, void shrinkage was observed in the films at short times, and void growth occurred at longer times.

INTRODUCTION

While single-phase thin films and epitaxial multilayer films have been widely studied, the effects of deposition method and post-deposition annealing on grain growth, stress and texture formation in thin films containing multiple phases have not been investigated in comparable detail. Films of this type have many potential applications ranging from magnetic memories to catalysts. The aim of our work is to expand the understanding of grain growth and coarsening in a binary alloy thin film.

Copper / silver alloys exhibit eutectic behavior at equilibrium, but rapid quenching from the liquid and vapor phases can produce γ' and γ'' metastable FCC solid solution phases at all compositions. These phases exhibit 1 – 3 % deviation from Vegard's Law, and can decompose into supersaturated α' (Ag-rich) and β' (Cu-rich) solid solution phases during annealing or long-term aging at room temperature [1, 2, 3, 4].

Previous works have been conducted with copper / silver alloy films. Stoering and Conrad studied electron-transparent regions in rapidly-cooled alloys and thermally evaporated copper / silver [~50 atom %] alloy films deposited on rocksalt substrates [3]. Supersaturated solid solution phases were observed, *via* electron diffraction and XRD, with grain sizes on the order of 10-20 nm. Since electron-transparent regions in the liquid-quenched alloys shrank and thickened due to surface tension when annealing temperatures exceeded 100°C, electron microscopy was limited to a maximum annealing condition of 100°C [3]. Boswell and Chadwick also examined liquid-quenched splat-cooled copper / silver [50 atom %] films on copper substrates. The γ' phase was present in XRD spectra for as-deposited films, and spectra for films aged four months at room temperature indicated complete decomposition to the equilibrium phases. Electron-transparent regions on the as-deposited films had grain size measurements from γ' - (a_o = 0.388 ± 0.006 nm) and γ'' - ($a_o \approx 0.400$ nm) dark field reflections varying from 3 to 2000 nm [4]. Barber studied copper / silver films deposited onto glass substrates and carbon-coated TEM grids heated at various temperatures during deposition, *via* dual magnetron sputtering of pure copper and pure silver. As-deposited films contained two crystalline FCC phases with predominant {111} textures. Texture strength decreased with increasing substrate temperature until no texture was apparent at 200°C. Deviations from the equilibrium lattice parameters were observed for the silver-rich and copper-rich phases formed on colder substrates, demonstrating an extended solid solubility for the two phases [5].

In this paper, we expand on the above works by depositing copper/silver alloy films, *via* PLD, and annealing these films for an extended length of time

MATERIALS and METHODS

Copper / silver alloy thin films were prepared by pulsed laser ablation of a custom-made target, and deposition of the vapor onto cleaved single crystal sodium chloride (rocksalt) substrates held at room temperature. The target was prepared by arc melting pure copper and pure silver to produce a eutectic composition of 66 Ag / 34 Cu [wt. %], cold-rolling the resultant slug, and annealing at 700°C for 24 hours. A Lambda Physik® LPX 300 KrF Excimer Laser of 248nm wavelength was used to ablate the target with pulses of 25 ns duration. The pulse frequency was 5 Hz for an overall deposition time of 15 minutes ± 1%.

For TEM analysis, films were separated from their rocksalt substrates using five dilutions of distilled water, and they were subsequently supported by 200-mesh copper TEM grids. The films were annealed atop their TEM grid supports, at 100°C, for 15, 30, 60, 180, and 1440 minutes, in a controlled atmosphere of 95Ar / 5 H_2 [vol. %] flowing at approximately 100 sccm. The heating ramp rate was 1000°C / hr, and a one-hour cooling time to room temperature was given after each anneal.

Each copper / silver alloy film sample was examined using a JEOL® JEM 2000 FX TEM. Bright field images, diffraction patterns, and axial dark field images corresponding to the copper/silver alloy {111} + {200} and {220} + {311} diffraction rings were collected and analyzed digitally with Gatan® Digital Micrograph. Pure gold film diffraction patterns were used for accurate camera-length calibration. The film compositions were measured using energy-dispersive x-ray spectrometry, and a Digital Instruments Multimode Atomic Force Microscope (AFM) was used in tapping mode to measure the thickness at the edge of an as-deposited film transferred to a glass substrate.

RESULTS and DATA ANALYSIS

The thickness of an as-deposited copper/silver film was measured with the AFM. The substrate was disolved in distilled water and the film was transplanted to a soda lime glass support. Height profiles were measured across the edges of film with thickness measurements ranging between 20 and 30 nm.

For each annealing condition, a series of bright field (BF) images was captured with the TEM, along with subsequent diffraction pattern and axial dark field (DF) images corresponding to the copper / silver alloy {111} + {200} and {220} + {311} diffraction rings. Figure 1 shows example BF, diffraction pattern and DF images for films as-deposited and annealed for 180 minutes and 1440 minutes at 100°C.

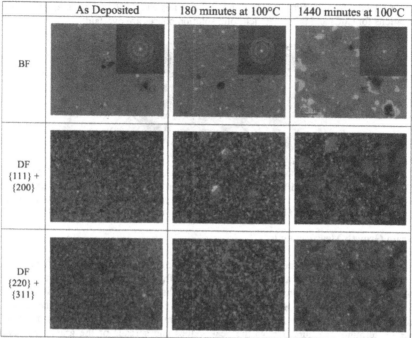

	As Deposited	180 minutes at 100°C	1440 minutes at 100°C
BF			
DF {111} + {200}			
DF {220} + {311}			

Figure 1: Bright field (BF), diffraction pattern, and dark field (DF) images for copper / silver films on a copper TEM grid as-deposited (left) and annealed for 180 minutes (center) and 1440 minutes (right) at 100°C.

In the TEM, the films produced FCC diffraction ring patterns with d-spacings that were slightly lower than those of pure silver, indicating a solid solution of silver and copper. On the solid solution diffraction pattern {200} ring, intense spots exhibiting four-fold symmetry that correspond to silver-rich {100} orientations are observed. These spots are more evident for annealed films, and they may be attributed to large crystals present in the matrix or a preponderance of small crystals that have a common orientation.

In addition, a complication to the experiment resulted in a secondary FCC phase corresponding to CuCl ($a_o = 0.541$ nm) or NaCl ($a_o = 0.564$ nm), which was evident in some of the diffraction patterns. While typical d-spacing measurements had a better fit with CuCl, small calibration uncertainties can allow attribution of the identity of this phase as the more plausible NaCl from the substrate. EDS spectra for such particles showed copper- and chlorine- rich compositions relative to the adjacent matrix; sodium K_α and K_β peaks were not discernable on typical spectra since their energies overlap with the energy of the copper L_α peak. The intensity of the diffraction pattern rings for this secondary phase coincided with the density of particles that are slightly darker than the matrix in bright field images of some of our specimens. This phase grew to large sizes over long annealing times, as shown in Figure 1. Figure 2 shows half of a diffraction pattern taken from a film annealed for 15 minutes at 100°C. The theoretical ring locations for CuCl and the silver-rich solid solution phases are overlaid.

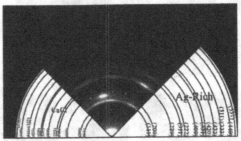

Figure 2: Cut view of a diffraction pattern taken from a copper/silver film annealed for 15 minutes at 100°C. The ring patterns for diffracted grains corresponding to CuCl and the silver-rich solid solution phase are overlaid.

Grain Growth Measurements

Grain sizes were measured for each annealing condition using dark field image sets for {111} + {200} and {220} + {311} diffracting grains and reported as the diameters of equivalent circular areas. Grain size distributions were obtained, and each was fitted to a lognormal distribution function. The peak value (often called the "mode" of the distribution) and the spread (± one standard deviation for the distribution function) were determined. Figure 3 plots the peak values and spreads of the grain sizes over annealing time for {111} + {200} and {220} + {311} diffracted grains, respectively.

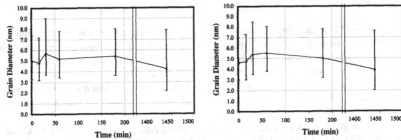

Figure 3: Lognormal peak values and spreads for {111} + {200} (left) and {220} + {311} (right) diffracted grains for copper / silver films annealed at the given times at 100°C.

Void sizes (measured as equivalent circular diameters) were measured and void area fractions were estimated from twenty bright field images with known frame sizes for each annealing condition. Figure 4 plots the mean void size with a normal distribution spread for films as-deposited and annealed for the five annealing times at 100°C. Table 1 gives the estimated void area fraction for each annealing condition.

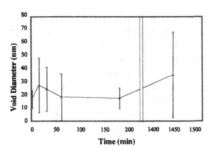

Figure 4: Void size distributions for copper / silver films annealed at the given times at 100°C.

	Estimated Void Fraction		Estimated Void Fraction
As Deposited	0.0056 ± 0.0023	60 min. at 100°C	0.045 ± 0.027
15 min. at 100°C	0.019 ± 0.0055	180 min. at 100°C	0.019 ± 0.0046
30 min. at 100°C	0.013 ± 0.0078	1440 min. at 100°C	0.097 ± 0.029

Table 1: Void size distributions for copper / silver films annealed at the given times at 100°C.

DISCUSSION

Based on dark field imaging analysis, a fine-grained silver-rich solid solution matrix was maintained even after annealing for 24 hours at 100°C. Separation into pure phases during annealing was apparent but minimal. Stoering & Conrad did not observe phase separation in as-deposited electron-transparent alloy regions. They also observed shrinkage in these regions when they were annealed at temperatures exceeding 100°C, preventing electron microscopy characterization [3]. Preliminary annealing experiments that we have performed on rocksalt, copper TEM grids, and Si_3N_4 at 200°C - 400°C verify that the films shrink and densify before phase separation can occur, even at these higher temperatures, in continuous films [6]. Boswell & Chadwick observed equilibrium phase formation in films aged for four months at room temperature [4]. In this work, the copper / silver solid solution phase was still present in diffraction patterns for films stored for as long as 50 days at 5°C and subsequently annealed for 15, 30, and 60 minutes at 100°C, although silver-rich clusters were observed in selected area diffraction patterns and CBED patterns. Particles that could be analyzed with EDS had copper- and chlorine-rich compositions relative to the adjacent matrix; no adequate spectra for silver-rich particles could be obtained. Due to uncertainty in the diffraction pattern measurements, a compositional distinction between a pure silver phase and the solid solution phase could not be made.

Minimal grain growth occurred in our films during annealing at 100°C, and indeed, grain shrinkage is arguably observed in Fig. 3. This may have been a result of the nucleation of new grains of successively more stable supersaturated phases.

For each anneal, the mean and lognormal peak grain sizes for {111} + {200} diffracting grains were similar to the mean and peak grain sizes for the {220} + {311} diffracting grains. Therefore, strong texture formation is not evident.

The mean void size grew after as-deposited films were annealed for 15 minutes at 100°C, but then shrank during subsequent anneals until it grew once again for the 24 hour anneal. If the driving force for void growth and shrinkage is surface energy reduction, then a stable minimum void size could be determined with a model, such as one derived by Srolovitz, et al [7]. The stable void size depends upon the void separation. Therefore, continued void nucleation during the annealing process can explain the rather unexpected changes in the sign of the growth rate.

Growth or shrinkage of voids requires long-range diffusion with transport distances roughly equivalent to half of the mean void spacing. Grain growth, in contrast, is a process that is mediated by short-range diffusion – single atomic jumps across a grain boundary, in the view of most models. It is, therefore, somewhat surprising that void growth is so much more pronounced than grain growth in our experiments, and we attribute this to the predominating effect of surface diffusion over all forms of bulk and grain boundary diffusion at the low temperatures that we have investigated thus far. The formation of second phase nuclei, and their subsequent growth, would require a combination of short-range and long-range diffusion processes.

The atomic mechanisms of diffusion in the supersaturated solid solution may be considerably complicated by atomic co-ordination and correlation effects, where the species have a tendency to separate. These effects may significantly reduce the bulk diffusivity in this phase.

CONCLUSIONS

Although thin films can undergo structural evolution by a number of diffusional processes, in Cu / Ag supersaturated solid solution films at 100°C void growth occurs while grain growth and transformation to the stable phases do not. These changes are dominated by processes that depend on long-range diffusion (most likely surface diffusion) rather than short-range diffusion through the bulk or the grain boundaries.

ACKNOWLEDGEMENT

This research project was supported by the National Science Foundation, grant number DMR-0504813

REFERENCES

1. P. R. Subramanian, D. J. Chakrabarti, and D. E. Laughlin, eds. *Phase Diagrams of Binary Copper Alloys*, ASM International: Materials Park, OH, 1994.
2. S. Nagakura, S. Tomaya, and S. Oketani, *Acta Metall.*, **14(1)**, 73-5 (1966).
3. R. Stoering and H. Conrad, *Acta Metall.*, **17(8)**, 933-48 (1969).
4. P. G. Boswell and G. A. Chadwick, *J. Mat. Sci.*, **12(9)**, 1879-94, (1977).
5. Z. H. Barber, *Vacuum*, **41(4-6)**, 1102-5, (1990).
6. S. T. Snyder, *Microstructural Stability In Copper / Silver Alloy Thin Films*, Purdue University (2005).
7. D. J. Srolovitz, W. Yang, and M. G. Goldinger, *Materials Research Society Symposium Proceedings*, **403**, 3-13, (1996).

AUTHOR INDEX

SUBJECT INDEX

Printed in the United States
By Bookmasters